FAMILY
engineering

An Activity & Event Planning Guide

Mia Jackson
David Heil
Joan Chadde
Neil Hutzler

Designer
Lauren Seyda

Illustrator
Keith Warner

Family Engineering Program Development Partner Organizations

FOUNDATION FOR FAMILY
SCIENCE & ENGINEERING

Developed with support from the National Science Foundation

For more information about the Family Engineering program or to order additional copies of *Family Engineering: An Activity and Event Planning Guide*, contact:

Foundation for Family Science and Engineering
4614 SW Kelly Avenue, Suite 100
Portland, Oregon 97239
Tel: (503) 245-2102
Fax: (503) 245-2628
www.familyengineering.org

Copyright © 2011 by Foundation for Family Science and Engineering and Michigan Technological University

Neither the publishers nor the authors shall be liable for any damage caused or sustained due to the use or misuse of ideas, activities, or materials featured in this book.

This book was written with the support of the National Science Foundation (NSF) Grant DRL-0741709. However, any opinions, findings, conclusions, or recommendations herein are those of the authors and do not necessarily reflect the views of the NSF.

Printed in the United States of America
ISBN: 978-0-615-49366-4
Library of Congress Control Number: 2011930777

10 9 8 7 6 5 4 3 2 1

Permissions
The five-step engineering design process used in the Family Engineering program is adapted with permission from *Engineering is Elementary®* *(EiE)*, Museum of Science, Boston, MA.

Tumbling Tower is adapted with permission from the *3-2-1 Contact Teacher's Guide, Season II*, ©1983 Sesame Workshop.

Who Engineered It? Activity Statements are adapted with permission from *Assessing Elementary School Student's Conceptions of Engineering And Technology* (Christine Cunningham, et. al., 2005).

Mining for Chocolate is adapted from the *National Science and Technology Week 1991 Teacher's Guide* with permission from Biological Sciences Curriculum Study (BSCS).

Cover Photos:
Front Cover: top, Capture Photography; middle, Craig Felix; bottom, David Heil
Back Cover: Ever After Photography and Production, LLC

About the Authors

Mia Jackson, an Associate with David Heil & Associates, Inc., specializes in program and exhibit development, project management, and evaluation with an emphasis on early learning, parent/child engagement, and public outreach. Trained in Elementary Education, she has worked in the science education field for over 15 years, including serving as the Director of Education and Exhibits at The Imaginarium, a science center in Anchorage, Alaska.

David Heil, President of David Heil & Associates, Inc., is well known as an innovative educator, author, and host of the Emmy-Award winning PBS science series, Newton's Apple. Active in promoting public understanding of science for over 30 years, he is a frequent conference and workshop presenter on science, technology, engineering, and mathematics (STEM) education. David was the lead editor of the popular book *Family Science* and the Founding President of the Foundation for Family Science and Engineering, one of the three partner organizations responsible for creating Family Engineering.

Joan Chadde is the Education Program Coordinator for the Center for Science and Environmental Outreach at Michigan Technological University (MTU), one of the three partner organizations responsible for creating Family Engineering. She has more than 25 years experience in science and environmental education, water resource management, and program development, including the design and implementation of numerous K-12 science programs and teacher professional development workshops.

Neil Hutzler, PhD., P.E., is a professor and past chair of the Department of Civil and Environmental Engineering and the director of the Center for Science and Environmental Outreach at Michigan Technological University (MTU). He has over 30 years of experience in engineering education in both K-12 and higher education. MTU is one of the three partner organizations responsible for creating Family Engineering and Neil served as the Principal Investigator of the National Science Foundation grant supporting the development of the Family Engineering program.

Preface

Collaboration is a hallmark of engineering and the creation of this publication, *Family Engineering: An Activity and Event Planning Guide*, is a perfect example of collaboration in action.

Since the release of their popular book *Family Science* in 1999, the Foundation for Family Science and Engineering (FFSE) has been promoting and supporting parents and other caregivers exploring science and engineering with their children.

Michigan Technological University (MTU) has been using engineering students to conduct Family Science nights at elementary schools and community settings across northern Michigan since 1997. With MTU's deep expertise in engineering, developing a family learning program focused on engineering was a logical next step and they contacted the FFSE to explore interest in collaborating on development of a new Family Engineering program.

As excitement for such a program grew, so did the collaboration. First, the FFSE tapped the consulting firm of David Heil & Associates, Inc. known for developing innovative, national scale science, technology, engineering, and mathematics (STEM) programs. Next, MTU invited the American Society for Engineering Education (ASEE) to join. ASEE had been expanding their own engineering education offerings targeted at K-12 teachers and students and saw the potential for Family Engineering to greatly increase public awareness and appreciation of engineering. Finally, the founding director of the innovative elementary classroom curricula, *Engineering Is Elementary (EiE)*, signed on as an advisor to the Family Engineering project team. Together, this group applied to the National Science Foundation (NSF) to support the research and development of a collection of activities and delivery formats suitable for actively engaging elementary-aged youth and their parents in exploring engineering outside of formal classroom settings. Once funded, the real fun began!

Developed collaboratively by MTU, the FFSE, and ASEE, the Family Engineering program is designed to introduce families to the excitement of engineering and the important contributions engineers make to the designed world we live in. With the NSF's support, the development team was able to pilot and field-test activities and different delivery formats in dozens of locations across the US and Puerto Rico, with volunteers ranging from classroom teachers, professional engineers, university students and faculty, museum educators, and parents. The result is this collection of innovative activities and resources for helping plan and host fun, highly interactive events for families to explore and discover engineering together.

While the program was under development, national attention on STEM education grew substantially, with numerous reports and initiatives pointing to the need to develop a 21st Century workforce and citizenry better skilled in problem solving, teamwork and, yes, collaboration. In your hands you hold a valuable tool to help accomplish this critical mission. *Family Engineering: An Activity and Event Planning Guide* was born out of collaborative teamwork, a passion for problem solving, and a shared commitment to making a difference in the world. These elements not only define the Family Engineering program and the hands on activities found in this publication, but they are also attributes that engineers exhibit every day in their work.

Enjoy!

Acknowledgements

We would like to thank the numerous education and engineering professionals who contributed to the development and field testing of activities and content for *Family Engineering: An Activity and Event Planning Guide*.

Project Advisors

Christine M. Cunningham, Ph.D., Founder and Director, Engineering is Elementary, Vice President, Museum of Science, Boston, MA

William E. Kelly, Ph.D., Manager of Public Affairs, American Society for Engineering Education (ASEE), Washington D.C.

National Advisory Committee

Sandra Begay-Campbell, Sandia National Laboratories, Albuquerque, NM

Gary Cruz, Ph.D., Great Minds in STEM, Los Angeles, CA

Deborah Curry, Discovery Place, Charlotte, NC

José Franco, WestEd, Oakland, CA

Anne Gurnee, Southwest Charter School, Portland, OR

Jason D. Lee, Detroit Area Pre-College Engineering Program, Detroit, MI

Paul Poledink, Ford Motor Company Fund, Dearborn, MI (retired)

Gerald Wheeler, Ph.D., National Science Teachers Association (Emeritus Executive Director), Arlington, VA

Content Contributors

Celeste Baine, Engineering Education Services Center, Seattle, WA

Gail Foster, Independent Consultant, Homosassa Springs, FL

Cathy Griswold, Mari-Linn School, Lyons, OR

Anne Gurnee, Southwest Charter School, Portland, OR

Elizabeth Parry, North Carolina State University College of Engineering, Raleigh, NC

Pamela Schmidt Bardenhagen, Michigan Technological University, Houghton, MI

Cary Sneider, Ph.D., Portland State University, Portland, OR

Expert Reviewers

Lydia Beall, Museum of Science, Boston, MA

Sandra Begay-Campbell, Sandia National Lab, Albuquerque, NM

Monica Cardella, Ph.D., Purdue University, West Lafayette, IN

Robert Claymier, The Children's Council of the International Technology and Engineering Educators Association, Reston, VA

Deborah Curry, Discovery Place, Charlotte, NC

Elizabeth Eschenbach, Ph.D., Humboldt State University, Arcata, CA

David W. Gatchell, Ph.D., Illinois Institute of Technology, Chicago, IL

Judith R. Hallinen, Carnegie Mellon University, Pittsburgh, PA

Karen Mayfield-Ingram, EQUALS, Lawrence Hall of Science, Berkeley, CA

Mercedes McKay, Stevens Institute of Technology, Hoboken, NJ

Susan Staffin Metz, Stevens Institute of Technology, Hoboken, NJ

Gerald Recktenwald, Ph.D., Portland State University, Portland, OR

Robert F. Vieth, Juneau Economic Development Council, Juneau, AK

Kristin Bethke Wendell, Ph.D., Tufts University, Medford, MA

Gerald Wheeler, Ph.D., National Science Teacher's Association (Emeritus Executive Director), Arlington, VA

Pilot and Field Test Site Coordinators

Grace Dávila Coates, EQUALS/Family Math Program, Berkeley, CA
Eric Heiselt, Mississippi State University, Mississippi State, MS
Krizia M. Karry, Industry University Research Consortium, San Juan, Puerto Rico
Marta Larson, University of Michigan School of Education, Ann Arbor, MI
Elizabeth Parry, North Carolina State University, Raleigh, NC
Jack Samuelson, Marquette University, Milwaukee, WI
Lara K. Smetana, Ph.D., Loyola University Chicago, Chicago, IL
Elizabeth Smith, Fort Discovery National Science Center, Augusta, GA
Ron Terry, Ph.D., Brigham Young University, Provo, UT

Family Engineering Pilot and Field Test Event Sites

Rosa Parks Elementary, Berkeley, CA
Kids' Breakfast Club, Hayward, CA
Richmond Library, Richmond, CA
Washington Elementary, Richmond, CA
TECH Museum, San Jose, CA
Wintergreen Interdistrict Magnet School, Hamden, CT
Casimir Pulaski School, Meriden, CT
Worthington Hooker School, New Haven, CT
Barnard Environmental Studies School, New Haven, CT
Fair Haven School, New Haven, CT
Episcopal Day School, Augusta, GA
Fort Discovery National Science Center, Augusta, GA
Lake Forest Hills Elementary, Augusta, GA
Blythe Elementary, Blythe, GA
Baraga Elementary School, Baraga, MI
Calumet-Laurium-Keweenaw Elementary School, Calumet, MI
Hannahville Indian School, Wilson, MI
Upper Peninsula Girl Scout STEM Extravaganza, Houghton, MI
Upper Peninsula Children's Museum, Marquette, MI
Golightly Education Center, Detroit, MI
Maybury Elementary, Detroit, MI
Roberto Clemente Academy, Detroit, MI
Washington Elementary, Flint, MI
Pinckney Elementary, Pinckney, MI

Brick Elementary, Ypsilanti, MI
East Webster Elementary, Mathiston, MS
Ward Stewart Elementary, Starkville, MS
Starkville Boys and Girls Club, Starkville, MS
Sudduth Elementary, Starkville, MS
West Oktibbeha County Elementary, Sturgis, MS
Brentwood Magnet School of Engineering, Raleigh, NC
St. Mary, Mother of the Church Catholic Parish, Garner, NC
Rachel Freeman School of Engineering Wilmington, NC
Mari-Linn School, Lyons, OR
Markham Elementary, Portland, OR
Southwest Charter School, Portland, OR
José Cordero Rosario School, Barceloneta, Puerto Rico
Luis Muñoz Marín School, Carolina, Puerto Rico
Ángela Calvani School, Jayuya, Puerto Rico
Rafael Quiñones Vidal School, San Juan, Puerto Rico
Manuel Padilla School, Vega Baja, Puerto Rico
Thanksgiving Point, Lehi, UT
Newington Forest Elementary School, Springfield, VA
Bruce Guadalupe Community School, Milwaukee, WI
French Immersion School, Milwaukee, WI
Milwaukee Academy of Science, Milwaukee, WI
Wilson School, Wauwatosa, WI
Waukesha STEM School, Waukesha, WI

Other Contributors

There are many different aspects involved in the development, design, and production of a publication. We would like to offer additional thanks to the individuals and organizations below for their contributions.

Gina Magharious, David Heil & Associates, Inc., Portland, OR
David Mitchell, American Society for Engineering Education, Washington D.C.
Kristina Peltier, David Heil & Associates, Inc., Portland, OR
Eric Wallinger, David Heil & Associates, Inc., Portland, OR
Lloyd Wescoat, Michigan Technological University, Houghton, MI
Amy Farrell, times2studio, LLC, Missoula, MT
Inverness Research Associates, Inverness, CA

Table of Contents

WHAT IS FAMILY ENGINEERING?

1

Family Engineering is an informal engineering education program that actively engages elementary-aged children and their families in fun, hands-on, engineering activities and events. The program is modeled after two other popular programs, each with their own published activities and event planning resources. The first is *Family Math* (1986), developed by the Lawrence Hall of Science, affiliated with the University of California, Berkeley (www.lawrencehallofscience.org/equals). The second is *Family Science* (1999), developed by the Foundation for Family Science & Engineering in collaboration with the Oregon Museum of Science and Industry (OMSI) and Portland State University, all based in Portland, Oregon (www.familyscience.org). This Family Engineering publication is full of exciting activities that use inexpensive, easy-to-find, materials, as well as event planning tools and resources to assist in hosting a fun-filled event for families in your community. The activities in this book can also be used to explore engineering at home.

Family involvement in Family Engineering is key to the program's success. Children may attend a Family Engineering event or engage in one or more of the activities included in this book with a parent, grandparent, older sibling, aunt, uncle, mentor, or other important adult in their lives. Their "family" consists of people close to them who are willing to support and encourage their personal development and education.

Family Engineering activities are designed for children aged 7-12 and their parents, or other adult caregivers, to be actively engaged as a team. Research shows a significant improvement in children's self-confidence and learning skills when families are more actively engaged in their learning. By showing interest and exploring engineering with their children, parents and other caregivers can positively influence a child's attitude about engineering, as well as encourage their children to consider a possible career in engineering.

Engineers play an essential role in our designed world, yet many people are unaware of what engineers actually do. We can change this for future generations by providing positive experiences with engineering at an early age. Increasing parents' and children's awareness, appreciation, and understanding of engineering will open their eyes to the significant impact of engineering in their daily lives, and to the tremendous career opportunities available in engineering around the globe. The workforce of tomorrow will need to be trained in science, technology, engineering, and math (STEM) in order to compete in the global market and contribute to society in the 21st century. Family Engineering addresses this challenge by introducing elementary-aged children and their families to the world of engineering and giving them an opportunity to explore and solve challenges together. A Family Engineering event can also be a chance for families to meet and interact with engineers or engineering students in an enjoyable, informal setting.

Research has found that free-choice learning experiences encourage a child's long-term interest in science, as well as significantly improving science understanding in populations typically underrepresented in science. Engineering, like science, needs to attract and engage a diverse population, including individuals from different cultures, genders, and backgrounds. Diversification of the engineering workforce will lead to better problem solving because broader perspectives will be applied to finding solutions. Family Engineering activities and events create opportunities outside of the formal classroom for children and their families to explore new and different types of learning, and, possibly, discover new interests and skills in themselves, and in each other.

Family Engineering Goals

* To engage families in engineering with fun, hands-on activities

* To increase public appreciation and understanding of the role engineering plays in everyday life

* To introduce children at an early age to the many career opportunities in engineering

* To increase parents' interest in and ability to encourage their children to pursue an engineering career

* To provide age-appropriate resources to support volunteers in conducting informal engineering education events with elementary-aged children and their parents

How to Use Family Engineering

Family Engineering: An Activity & Event Planning Guide is written to encourage, support, and assist with the organization and implementation of community-based Family Engineering events. These events may be held on weekday evenings or weekends in schools, community centers, faith-based facilities, or local businesses interested in promoting engineering education. Events may be organized by parents, classroom teachers, professional engineers, college students, museum educators, scout leaders, and other family organization leaders willing to volunteer their time to plan and host Family Engineering in their community. The activities in this book can also be used by families to explore engineering at home.

Specialized knowledge in engineering and/or years of teaching experience are not necessary to enjoy Family Engineering activities or events. A Family Engineering event provides an accessible and friendly informal learning environment that makes families feel comfortable and confident exploring new topics and approaching challenges together. Family Engineering activities invite families to play around, experiment, and try things out, without worrying about having the right answer.

Explore *Family Engineering: An Activity and Event Planning Guide* and find out how you can open up the world of engineering to children and families in your community. For more information about Family Engineering and additional engineering and family learning resources, visit **www.familyengineering.org**.

FAMILY LEARNING

2

Research has shown that families can have a major influence on a child's achievement in school and in life. Parents, and other adult caregivers, can make a significant difference when they become actively involved in their children's education. They can create a positive environment for learning by encouraging, supporting, and even participating as their children ask questions and explore new interests.

The design, approach, and materials used in Family Engineering activities encourage family interaction and invite families to work together as a team. Family Engineering events create an environment that promotes collaborative family learning. When organizing and implementing a Family Engineering event, there are a number of ways to help foster a successful family learning environment.

Supporting Family Learning at a Family Engineering Event

Set clear expectations

Make it clear on event flyers that the event is for adult family members and children *together*. When doing follow-up phone calls, emails, or notices, remind families that parents, or other adult caregivers, will be engaging in activities *with their children*.

Prepare event volunteers

Give volunteers information and tips on how to encourage and support family interaction. Provide them with the *Working With Families: Tips for Event Volunteers* handout available in Appendix E.

Set the stage

As families arrive at an event, remind them to stay together as a family and to engage in the activities as a team.

Promote family interaction and involvement

Invite the whole family to work together by encouraging adults to join the children in doing each of the activities. Encourage families to involve every family member when participating in an activity.

Allow families to explore and discover on their own

Try not to show or tell families how to complete an activity. Use encouragement and questions to help them explore and experiment on their own.

Model a healthy attitude toward learning

Show an enthusiastic attitude towards engineering, problem solving, and accepting fun challenges. Encourage families to try activities and use effective questioning skills to promote collaborative exploration and thoughtful interaction.

Recognize parents

Thank parents for taking the time to attend the event with their children. Let them know that their involvement makes a difference in their children's attitudes, interests, and achievements in school and in life.

Encourage continued learning at home

Close the event with suggestions for exploring engineering at home, and distribute the *Exploring Engineering at Home* handout available in Appendix E.

Encouraging Engineering Exploration at Home

Parents and other adult caregivers can support their children's continued exploration and interest in engineering at home and in their daily life by following these simple tips. To share this information with families at a Family Engineering event, distribute the handout *Exploring Engineering at Home: Tips for Parents and Other Adult Caregivers* available in Appendix E.

Encourage and support an interest in engineering

By showing an interest in engineering yourself, you'll build a positive attitude toward engineering in your children. Help your children explore engineering on the Internet; find books or magazines about science and engineering at the library; watch movies or television programs that involve design and engineering; or try their hand at designing and building an invention of their own. Identify engineers working in your community and invite them to be guest speakers at your children's school, or arrange a field trip to the engineer's workplace. Try out some activities from *Family Engineering* at home.

Model problem-solving strategies

Children who learn to work their way through problems and explore different solutions become more capable and confident problem-solvers as adults. Show your children different strategies for solving problems:

- draw a picture or diagram
- talk it over with a friend or family member
- find an expert to offer advice
- break the problem down into smaller pieces
- brainstorm multiple options or approaches to a solution
- design and build a prototype for testing

Point out when you are using these strategies yourself and find opportunities to solve problems together as a family. Encourage children to look at their everyday lives and identify problems that need solving or ways of doing things that could be improved. Acknowledge when a solution or a design fails or doesn't work out the way you planned. Talk to your children about what didn't work and how you can use this experience to change something to make it work better in the future.

Encourage questions

Children possess a natural curiosity about the world and are motivated to seek answers to their own questions. Teach them how to do this by reading books, using the Internet, experimenting, or taking things apart. Don't worry if you don't know the answers. Sharing in the joy of discovery is a wonderful model for lifelong learning. Encourage children to ask questions by posing questions yourself.

- *What happened?*
- *What should we try next?*
- *What will happen if?*
- *Can you show/tell me how this works?*

Help children "do it themselves"

Try not to give your children a solution or answer to a problem. Instead encourage them to keep trying. Guide them with questions, hints, or clues. Let them make mistakes and correct themselves. Children who learn to work their way through problems, explore different solutions, and who learn to overcome their fear of failure become more capable and confident learners.

Support science and mathematics learning

Engineers use science and math knowledge and skills regularly in their work. Model a "can-do" attitude about science and math, recognizing the ways they are useful in daily life, such as counting change, predicting the weather, cooking, and fixing a flat tire on a bike. You do not need to be a science or math expert to encourage your children's success in these areas. Help them develop good study habits and show an interest in what they are learning in school. Encourage participation in science and math activities outside of school, such as clubs, competitions, after-school programs, or museum classes.

Recognize the creative side of engineering

Engineering is a creative endeavor. Combining objects in new ways, producing new uses for objects, solving problems and puzzles, pretending, dreaming, designing, and inventing things are all a part of engineering. Encourage your children's imaginations by providing open-ended toys, creative materials, and a safe place that allows them to design their own experiences, dream up imaginary worlds, invent interesting products, or design innovative structures.

Demonstrate how engineering improves the way we live

Engineering is all around us! Next time you are walking or driving around town, shopping in a mall, or just hanging out at home, look around with your child and see how many items you can find that were designed by engineers to solve a problem or meet a human need. Help children see how engineered products are a part of everyday life, and that engineers design things that improve the way we live.

Challenge stereotypes about who does engineering

Meeting or learning about female engineers or engineers from diverse cultures and backgrounds will demonstrate to children that a variety of people can be engineers. This allows children to imagine themselves as engineers. In addition, learning about the wide diversity of engineering fields and potential careers can help children link their own interests and skills to engineering.

FAMILY ENGINEERING AND EDUCATION STANDARDS

3

Countries around the globe use educational standards to describe what students should know and be able to do prior to emerging from their pre-college school experience. The content-knowledge and skills described in these standards are used to guide decisions about curriculum, instruction, and assessment suitable for each grade-level as a learner progresses through their formal K-12 education.

In the United States, as well as many other countries, educational standards have been adopted to promote scientific, technological, and mathematical literacy. At the time that this publication went to press in 2011, the United States had just completed development of new Common Core State Standards in Mathematics, and was initiating the development of Next Generation Science Education Standards. Based on recommendations from the National Academy of Engineers, the National Research Council, the National Science Teachers Association, and numerous professional associations and individuals engaged in K-12 education, these Next Generation Science Education Standards include engineering ideas, concepts, and skills considered important to a student's understanding of how science is applied, and the inter-relationships between science, technology, engineering, and society.

Family Engineering has been designed to introduce families to many of the concepts and skills found in national science, technology, and mathematics standards. In particular, Family Engineering activities focus on the engineering concepts and skills infused in these standards, including the engineering design process. While not developed to be used as a formal classroom curriculum, Family Engineering can still be a valuable resource for educators, helping them provide students, and their families, with relevant experiences applying science and mathematics through engineering. In addition, Family Engineering provides interested individuals from outside the formal K-12 school system with creative activities and resources that complement and reinforce many of the concepts and skills being taught in school. These individuals include parents, professional scientists and engineers, college students and faculty, and informal educators from museums and science centers. This allows for broad community-based engagement in science, technology, engineering, and mathematics (STEM) learning, an important ingredient in cultivating a scientifically and technologically literate citizenry for the 21st century.

In addition to complementing specific content standards in mathematics, physical, earth, and life sciences, Family Engineering activities and events also provide families with hands-on experiences in many of the skills and practices outlined in the standards documents. While identified as essential to scientific and technological literacy for all, these skills and practices are also used on a daily basis by trained scientists and engineers when conducting investigations and designing solutions to real-world problems. As a result, Family Engineering helps all families better understand the designed world. It also introduces parents and elementary-aged children to a range of practical and important skills necessary for a career in a science or engineering field.

> *Family Engineering helps families better understand the designed world. It also introduces parents and elementary-aged children to a range of practical and important skills necessary for a career in a science or engineering field.*

In particular, here are a few of the science, technology, engineering, and mathematics standards, practices, and skills addressed in Family Engineering:

- Asking questions and seeking answers to these questions

- Making sense of a problem and designing solutions to address the problem

- Developing and using models

- Planning and carrying out experiments and fair tests

- Making observations, evaluating data, and communicating findings

- Using a design process to propose or improve a solution

- Using appropriate tools and techniques to design, build, and test an innovation

- Understanding the relationships between science, technology, engineering, society, and the natural world

For a full list of the engineering concepts and skills infused throughout the Family Engineering program, turn to page 13 in chapter 4, "The World of Engineering."

THE WORLD OF ENGINEERING

Engineers design, create, imagine, innovate, and invent. They identify and analyze problems or needs in the world, and then search for ways to meet these challenges with new solutions. The products of engineering are all around us. From the things we buy to the services we use, we are touched by engineering every day. Products in our home, such as televisions, computers, packaged foods, band aids, can openers, furniture, windows, and heating systems, as well as many aspects of our communities, such as buildings, transportation systems, electricity, airports, and waste disposal systems, are all engineered.

Engineers create useful tools and products that solve everyday problems. But they also work on large-scale challenges, such as designing the treatment and delivery of safe drinking water around the globe, designing new medical technology to help the sick and injured, designing ways to use energy more efficiently, and designing buildings that can survive major earthquakes and storms. The world of engineering is truly the world in which we live.

This section describes how engineers do their work using an engineering design process. It also lists and defines some common fields of engineering, and highlights a number of concepts and skills that are important in engineering.

Engineering Design Process

Engineers use their knowledge of math and science, along with engineering principles and experience, to create solutions to problems. Because engineers do this on a regular basis, they have developed a problem-solving tool called the **engineering design process** to guide their work. The design process can be used to approach just about any problem. It is a way to explore different ideas, try out possible solutions, learn from mistakes, and continually improve a product or process.

The engineering design process is cyclical, continuing until an effective design is created. It is not a rigid, step-by-step process. Engineers may start at any one of the stages and move back and forth between them as they work through a solution. For example, an engineer might start with the 'Improve' step when identifying an existing product or process that needs to work more effectively, or start at the 'Ask' step when confronted with a new problem. Often, engineers work in teams. One group of engineers might be responsible for one of the steps, and then pass the project off to another group to work on the next step.

The engineering design process can be simplified and presented as five steps that are easy for elementary-aged children and their parents to understand. Family Engineering has adapted the five-step engineering design process developed by *Engineering is Elementary*® (EiE), an innovative elementary classroom curriculum created by the Museum of Science in Boston, MA.

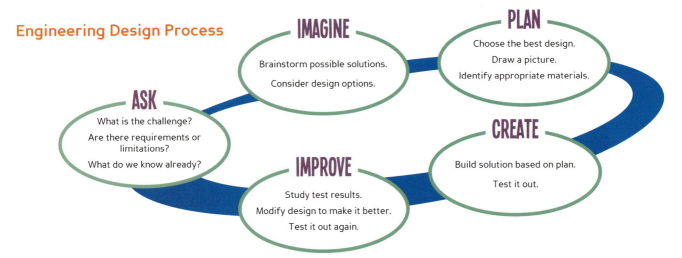

Engineering Design Process

IMAGINE
Brainstorm possible solutions.
Consider design options.

PLAN
Choose the best design.
Draw a picture.
Identify appropriate materials.

ASK
What is the challenge?
Are there requirements or limitations?
What do we know already?

CREATE
Build solution based on plan.
Test it out.

IMPROVE
Study test results.
Modify design to make it better.
Test it out again.

When engineers are confronted with a new challenge, they **ask** questions. What is the nature of the problem? Who is affected by the problem? What are the limitations in terms of cost or available time to work on a solution? Does the end-product need to be a certain size or weight, or be made from a specific material? Once engineers have defined the nature of a particular problem, and the context within which it must be solved, they can actually begin to solve the problem itself.

Engineers think creatively as they brainstorm and **imagine** solutions. There are usually many ways to solve a problem or meet a challenge. This is where an engineer's experience is helpful. In addition, they may ask other people with different skills and experiences to suggest solutions.

Once a tentative solution is selected, engineers develop a **plan** for how to design and build their solution. Diagrams are made, materials are specified, and a set of directions are created. Often, there are multiple solutions to a problem. When this happens, the solutions are compared to one another, to see which one best solves the problem.

Next, engineers **create** their solution and test it to see how well their idea works to solve the problem. Lastly, engineers analyze their test results, share their work with others, and look for ways to **improve** the design. This may mean repeating the engineering design process until everyone is satisfied with the solution. For many projects, engineers from multiple fields work together to solve a problem.

Many of the Engineering Challenge activities in Family Engineering highlight the engineering design process. At the appropriate time during an activity, usually before distributing materials, facilitators should introduce families to the engineering design process and briefly review how they can use the design process during the activity. Then, while families are engaged in the activity, point out places where they use a step of the design process, or remind them that using the design process may help them find a solution to their challenge. A full-page version of the engineering design process is available in Appendix E and can be copied and posted, or distributed to families at a Family Engineering event.

Common Engineering Fields

Because there are so many different types of problems, there are many different types of engineers. Some engineers design medical equipment and tools, while others design computers, food preservatives, or new packaging materials. Often, different types of engineers work together on a team to solve a problem, or design a product or process. One goal of Family Engineering is to introduce families to the wide variety of engineering fields and potential career paths among those fields.

Different types of engineers work together on a team to solve a problem, or design a product or process.

Each Family Engineering activity lists the engineering fields related to that activity. For your reference, these different fields of engineering are listed below. There are many more fields and sub-fields of engineering to be explored, each with their own training and career pathways. This list can be a starting point for discovering the vast and diverse world of engineering.

Acoustical engineering uses the science of sound and vibration (acoustics) to design products that manipulate or control sound. These engineers often work to reduce unwanted sounds, such as noisy traffic, or the transfer of sounds through the walls of offices, hotel rooms, or recording studios.

Aerospace engineering is the design, testing, and building of aircraft and spacecraft, such as airplanes, helicopters, and rockets.

Biomedical engineering focuses on solutions to health problems and injuries. Biomedical engineers design mechanical devices and new technologies such as surgical equipment and artificial limbs.

Chemical engineering uses basic science and design to turn raw ingredients into useful products, such as food, medicines, and fuels. Chemical engineers also design the processes and the large-scale chemical production facilities to safely and reliably produce these products.

Civil engineering is used to design and build things like bridges, roads, airports, dams, and buildings. These engineers often specialize further in a specific area, such as transportation or large-building construction.

Computer engineering deals with the design of different parts of computers, such as memory chips, circuits, and sensors, as well as how computer systems function.

Electrical engineering uses electricity and electronics to design and build useful products, such as cell phones and televisions. Electrical engineers also design ways to get electricity from its source to the places that use it, for example, homes, factories, and businesses.

Environmental engineering focuses on protecting and improving the environment and human health. Environmental engineers design products and systems for preventing or cleaning up pollution; they design systems for recycling and waste disposal; and they develop procedures for monitoring air and water quality.

Geological engineering looks at soil composition and rock formations to analyze stability and guide where, and how, structures should be built in order to withstand earthquakes, avoid ground water, and provide stable, safe foundations. Geological engineers also help locate and measure mineral and energy resources that are hidden underground.

Industrial engineering helps people work in more efficient ways by organizing materials, machines, and information to optimize systems. Industrial engineers may design new processes or help improve existing processes, such as scheduling airline flights, distributing relief aid in a disaster, or managing crowds at an amusement park. They may design methods to make a high-quality product in a safe environment, while also striving to save time, money, materials, or energy.

Manufacturing engineering helps people make and distribute new products. Manufacturing engineers are involved in every step of the manufacturing process, from the initial design of a product, to creating the machines and systems that make the product. They also design the way the product is packaged and shipped to customers.

Materials engineering utilizes knowledge about the properties of different materials to design new materials, combine materials, or choose the best materials to make products that meet a specific need.

Mechanical engineering deals with designing, producing, and maintaining mechanical devices. These engineers work with all kinds of machines, from engines, to power plants, to robots. In addition, they design new machines to particular specifications.

Mining engineering designs methods to remove minerals found in the environment. Mining engineers use science and technology to figure out the best way to remove earth minerals with minimal harm to the environment.

Package engineering is the design of packages for products such as food, toys, and electronics. Package engineers design packages that protect a product, are appealing to customers, and are cost-effective to produce and ship.

Engineers design, create, imagine, innovate, and invent. They identify and analyze problems or needs in the world, and then search for ways to meet these challenges with new solutions.

Safety engineering utilizes engineering knowledge and skills to make sure that the design of a system, product, or process is safe and does not pose a risk of injury or property damage. Safety engineers work to eliminate unsafe practices and conditions in the workplace, analyze new products or methods for safety, and create safety guidelines and recommendations.

Systems engineering. When there is a large, complex project, systems engineers help people work together. They think of ways to use machines to make the work easier, and organize and monitor all aspects of the project.

Essential Concepts and Skills in Engineering

Through education and experience, engineers learn a number of concepts and skills. Science and mathematics knowledge are important aspects of engineering. However, additional skills, such as problem solving, teamwork, and communication, help engineers to be successful, and are useful to other, non-engineering professionals as well. Family Engineering has identified the following list of "Essential Concepts & Skills in Engineering" and incorporated these into the Family Engineering activities found in this book.

Communication: the ability to write, speak, listen, visualize, and problem-solve. An engineer must be able to share his or her work with co-workers, clients, regulators, and the public.

Controlled experimentation and testing: the ability to design and conduct experiments, as well as analyze and interpret data. Engineers must have an understanding of what constitutes a "fair" test.

Design requirements: When engineers are designing a product or process, they are given *design requirements,* which state the objectives for the final product or process (what it should be or be able to do). *Design specifications* are statements that explain the rules for how to make the product or process. In the Family Engineering activities, the term *design requirements* is used to describe both the requirements and specifications, together.

Engineering under constraints: the ability to design, manufacture, or build a component, product, or system to meet a particular outcome or solution while accommodating a range of constraints. Constraints may be economic, environmental, social, political, ethical, health, safety, or materials-related.

Engineering design process: a series of steps that engineers use to guide their problem-solving. Family Engineering uses a simple 5-step version of the engineering design process adapted from the one developed by the Museum of Science in Boston, MA for their *Engineering is Elementary*® (EiE) classroom curriculum. The steps are: Ask, Imagine, Plan, Create, and Improve.

Invention/innovation: the creation of new products or the improvement of existing products. Engineers are always looking for creative new ways to make, combine, or re-purpose things to meet a new need or solve a problem.

Modeling: the recognition that designs can often be optimized by building physical, visual, or mathematical models, and/or constructing and testing prototypes, prior to building a final product or completing a project.

> *Through education and experience, engineers learn a number of concepts and skills.*

Open-ended problem-solving: the ability to identify, formulate, and solve problems with more than one possible solution.

Optimization/trade-offs: the ability to get the most out of a process, or make trade-offs in order to enhance benefits and minimize negative impacts.

Properties of materials: the specific characteristics of a material that distinguish it from other materials, and/or determine how that material will perform in certain conditions, or contribute to a particular solution.

Role of failure: the recognition that failure plays an important role in the design process and is not necessarily a negative outcome. Engineers study failures to find better solutions.

Reverse engineering: the deconstructing, or taking apart of a product, process, or project to figure out how it works.

Sustainability: the capacity of a system to endure. Sustainability ensures that a biological system remains diverse and productive over time. In our modern world, engineering designs are increasingly expected to enhance our lives while meeting environmental standards, as well as providing for economic stability or growth.

Systems: a set of interconnected parts that work together to form an integrated whole.

Spatial visualization: the ability to envision 3-dimensional models from 2-dimensional drawings, and to "see" how things fit together.

Teamwork: working with a group of people, sometimes with different but complementary skills, to complete a task or project.

ORGANIZING A FAMILY ENGINEERING EVENT

5

Family Engineering is an exciting program that brings adults and children together to discover, explore, and enjoy the world of engineering. While any of the activities in this book may be done at home or with small groups of families, such as at a birthday party, neighborhood picnic, or a scout troop meeting, activities may also be used at events focused specifically on the Family Engineering program and experience.

A Family Engineering event creates an informal, fun environment that introduces engineering and engineering careers to parents and their children, specifically elementary-aged children. It can be a special evening event at a local school, a Saturday event at a neighborhood community center, or even a series of events where families attend three or more sessions over multiple weeks.

The key to a Family Engineering event is the FAMILY—children and their parents, or other caregivers, interacting and learning together. This is not an event where the children participate in activities while the adults stand back and watch or stay home. The parents are an essential part of the action!

Anyone who wants to get parents and children interested and excited about engineering—teachers; parents; professional engineers; college students majoring in engineering, science, or education; museum educators; scout leaders; and more—can host a Family Engineering event. Schools may want to use a Family Engineering event to encourage and enhance parent involvement. Engineering societies or companies that employ professional engineers might offer Family Engineering events in their community to encourage more children, and their parents, to consider engineering as a future career. Universities may enlist undergraduate science and engineering students to conduct Family Engineering events as an educational outreach offering in their local community.

Family Engineering activities are easy to facilitate. They do not require specialized knowledge of engineering or years of teaching experience. Activity descriptions provide detailed steps for leading families through the experience, from how to introduce engineering concepts and make real-world connections with the activity topic to how to distribute materials and effectively manage a gathering of families.

Family Engineering activities are easy to facilitate.

The Family Engineering website, **www.familyengineering.org**, provides additional information and resources useful for planning a successful event.

Getting Started

Form a Planning Committee

Planning a Family Engineering event is more fun and less work when you have plenty of help. Before getting started, brainstorm a list of people who could help, such as teachers, parents, or colleagues who may be willing to join a planning committee. If the event is a collaboration between two or more organizations, such as a local business, university, or professional engineering society and an elementary school, it is helpful to include representatives from all partner-organizations in the planning process, as well as during the actual event itself. If people feel as if they have been part of the planning process, they will be more likely to help promote the event, recruit others to volunteer, and participate fully themselves. The planning committee can work together to select specific activities for the event, and then carry out the necessary tasks to make it happen. Use the *Family Engineering Event Planning Checklist* (Appendix E) as a guide for organizing and implementing your event.

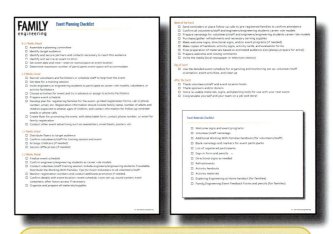

Family Engineering Event Planning Checklist available in Appendix E.

Determine the Target Audience and Event Size

An important initial planning step is to decide who will be invited to attend. Family Engineering events are generally designed for elementary-aged children and their parents. The activities are targeted for children between the ages of 7-12, but older or younger children can still enjoy participating as part of a family group.

The key to a Family Engineering event is the FAMILY.

Remember that "family" is the key element of a Family Engineering event. For a school-based event, invited participants may come from a single classroom, grade level, or the entire school, depending on the size of the event space and number of available volunteers. In a more open, community-based event it will be important to clearly communicate the appropriate age-range of children who may attend, as well as the fact that they must be accompanied by at least one parent/caregiver. For most Family Engineering events, keeping the maximum number of participants (children plus parents/caregivers) between 100-150 will ensure that every family has a quality experience.

Choose an Event Format

Do some initial thinking, based on your target audience of families, about how you will structure your event and which activities you would like to include. A typical Family Engineering event invites families to participate in 1 ½-2 hours of hands-on engineering fun. As families arrive, they immediately engage in a number of short (3-5 minutes), self-directed, tabletop activities called Openers. Openers consist of a few materials and a sign set out on a table. Families can choose to do as many Opener activities as they have time for, working at their own pace. Next, families sit at tables to participate in longer, facilitated Engineering Challenges. These longer activities, ranging in length from 20-50 minutes, can be facilitated for one large group of families or by using multiple activity facilitators with smaller groups of families in separate rooms. Listed below are a few possible formats for a Family Engineering event. See pages 30-31 for detailed *Sample Event Schedules*.

Single Large Group Format

This event format works best if you have a large space such as a school gym or cafeteria in which to host the event, as well as one or two lead facilitators that are comfortable in front of large groups. This format fosters community spirit as the group has a common experience and tackles the engineering challenges together. It also allows for less time to be spent moving families to separate rooms or changing locations in between activities.

Opener Activities—All event participants engage in the Opener activities at the same time. Having the Opener activities available for exploration at the beginning of an event gives every family a chance to be actively engaged the moment they arrive.

Engineering Challenges—Following the short Opener activities, the families remain together in one large group for a welcome and orientation to the event. Then, while still seated, they participate in 2-3 Engineering Challenges. One or two lead facilitators welcome the group, facilitate the activities, and wrap up the event. Other event volunteers assist with materials distribution and act as resources for families by circulating amongst the tables to encourage exploration and answer questions.

Sample Event Schedules available on pages 30-31.

Multiple Smaller Groups Format

The multiple smaller groups format works well if only small spaces are available, or when facilitators are more comfortable presenting to a smaller group of participants. This format allows activity facilitators to work with smaller groups of families, and for families to have a more intimate, focused experience. It can also allow you to offer a wider variety of activities during a single event.

Opener Activities—In this format, you may have all event participants engage in the Opener activities at the same time, or the Opener activities may be offered as one of the rotations for smaller groups of families to engage in at separate times during the event.

Engineering Challenges—Upon arrival, families are divided into groups of 20-25 participants and assigned to smaller rooms to engage in the Engineering Challenges. Families may be directed first either to a large area for Opener activities or to their separate activity rooms. A variety of different activities may be offered simultaneously, or the same activity may be conducted concurrently in multiple rooms. In each room, an activity facilitator leads the activity. If available, an activity assistant may be asked to help with materials distribution and act as a resource for families by circulating amongst the tables to encourage exploration and answer questions. One facilitator may lead several different activities for the same small group of families, or families may change rooms after each activity, allowing individual facilitators to focus on facilitating a single activity several times.

Series of Family Engineering Events

In this format, families sign up for several 1-2 hour events, offered once per week for several weeks. For example, you might offer a Saturday morning session from 9:30-11am for four consecutive Saturdays. Each event begins with a selection of Opener activities and then the group engages in 1-2 different Engineering Challenges. This format lends itself to a smaller group of families being led by a single activity facilitator and a small number of event volunteers. It allows for deeper engagement with families over time, and a chance to reinforce concepts and messages about engineering in a variety of ways with the same group.

Family Engineering at a Fair, Festival, or Exhibit Hall

Venues such as a county fair, Engineer's Week event, or a special community festival may serve families by offering a variety of self-paced, open-exploration experiences set up on tables or in an exhibit booth. Family Engineering Opener activities are an excellent resource for these types of family events because they are fun and engaging, yet easy to set up using simple materials. Opener activities do not need to be directly facilitated and only require that someone be available for questions and to invite families to become engaged. While this is not a full-fledged Family Engineering event, it can still serve to stimulate interest in engineering amongst family members. In addition, it may be used to recruit families for an upcoming Family Engineering event.

Choose a Location

Family Engineering events may take place in a variety of locations, such as school cafeterias or multi-purpose rooms, gymnasiums, community centers, church fellowship halls, libraries, museums, or even at home. Most importantly, the event location should be comfortable and convenient for families. Choose a location that provides ample parking, access by public transportation, and is safe and well lit for evening events. It should also have adequate restroom facilities and be accessible for handicapped individuals.

The facility should provide a pleasant learning atmosphere and have adequate space for the event, as well as enough tables and chairs for all participants and activities. Consider any special needs for the event—will families be engaged in large or small group settings, will a sound system be needed, or will supervised childcare for very young children be provided in a separate area?

Schedule the Event

Family Engineering event schedules can vary, but generally the most successful events are 1 ½-2 hours long, or a bit longer if a meal is served as part of the event. Choosing a "family friendly" time is very important. It is often successful to hold an event in the early evening on a weeknight, allowing time for parents to get home from work and gather the family before heading out to the event. Depending on your location and the families who will be attending, a Saturday morning event may also work well.

Make sure the location you have in mind is available for your chosen date and time, allowing adequate time before and after the event for setup and cleanup. Schedule your Family Engineering event far enough in advance (8-12 weeks) so that you have adequate planning time, and your target audience has time to find out about the event and register in advance.

Get Support

Although Family Engineering uses simple, inexpensive materials, there will still be some costs involved in putting on a Family Engineering event. These costs include activity materials, nametags, signs, and any refreshments that you plan to provide. There may also be some costs associated with the facility you are using for the event.

Don't be afraid to ask for donations. Many people will be excited about the idea of a Family Engineering event in their community and may be willing to contribute materials, food, time, or even money. Consider approaching local businesses, professional engineering societies, philanthropic organizations, a school's parent/teacher organization, or neighborhood associations. You can also request funding or submit grant proposals to the school district, the State Department of Education, or local foundations.

When asking an individual or group for support, be clear and concise. Explain what event you are planning and why you think that Family Engineering is important for your community. Outline exactly what you need (money, donated supplies, etc.) and how this individual/group can help. Offer to acknowledge their support on an event flyer or promotional poster, during the event, and/or in any media releases you distribute. Following the event, don't forget to send a heartfelt thank you to all your supporters.

Getting the Word Out

Inviting Families to Attend the Event

Once you have decided on the target audience and the number of participants your event can accommodate, the next step is inviting families to attend. Choosing a convenient date, time, and location are important factors in helping families decide to attend. You may also want to offer incentives to participate.

To help you prepare and plan for the Family Engineering event, it is recommended that you ask families to sign up in advance. You can do this by including a detachable registration form to an event flyer (see *Family Engineering Event Flyer* in Appendix E) or including a return phone number or email address on the invitation. This lets you know how many families to expect, and also allows you to do follow-up reminder phone calls or emails a few days before the event to families that have pre-registered.

Publicizing a Family Engineering Event

Think about publicity in the early planning stages, so you have plenty of time to create materials and distribute them in advance of your event. Once families know that a Family Engineering event is offered in their community, chances are that they will want to attend!

Tips for Motivating Families to Attend an Event

* Serve food (snacks or a meal)
* Offer raffle prizes or fun take home activities
* Provide baby sitting for very young children
* Invite an interesting or well-known special guest
* Advertise that the event is FREE
* Emphasize the hours of quality family time participants will enjoy
* Encourage teachers to give extra credit to their students who attend
* Collaborate with other organizations to encourage participation, such as scout troops, 4-H chapters, church groups, or youth clubs

Key Messages

- **Who** is it for? (an event for the entire family, targeting children 7-12 and their parents/caregivers)
- **What** is the event? (an educational and fun-filled evening or Saturday morning of hands-on activities)
- **When** is it being held? (day of the week, date, time)
- **Where** is it being held? (name of building, street address, city)
- **Why** should people come? (for quality family time, to learn about engineering, to do hands-on activities, etc.)
- **How** to sign up? (return a registration form, call or email the event coordinator, etc.)
- **It's FREE!** (or, if there is a fee, make sure that you state it)
- **Incentives** (mention if you are serving refreshments or if there will be prizes, special guests, or other incentives for attendance)
- **Sponsors** (be sure to recognize any sponsors in all advance publicity)

Types of Publicity

An event designed for a small group of neighborhood families or members of a specific organization may only require that a flyer be sent home or an invitation be extended through email or phone calls. If you are trying to reach a larger audience, however, you may want to distribute flyers more widely, put up posters, send out email notices, or notify local media outlets.

Flyers—Send Family Engineering event flyers home with students at a school; send them to civic and youth organizations to distribute to members; post on bulletin boards or enclose in a monthly newsletter; or place a stack in the library, local grocery store, or coffee shop. Include a section on the flyer that can be filled out and returned to register for the event, or instructions for registering by phone, email, or website.

Posters—You may want to create some large, colorful posters to hang in schools, grocery stores, churches, libraries, etc. Think of places where families from your target audience will most likely see the posters. A stack of event flyers can be placed next to a poster for interested families to take home.

Email Invitations—If you have access to an email list of families, send an electronic invitation with an option for families to register by responding directly to the email.

> Sample *Family Engineering Event Flyer* available in Appendix E.

Personal Invitations—Never underestimate the power of direct contact! Recruit volunteers, or assign members of your event planning committee to personally call or invite at least 5 families they know. If you are planning a smaller scale event for a very targeted audience, you may want to consider personally inviting each family with a phone call, email, or a mailed invitation.

Spread the Word—Provide event information to school or community organizations that frequently distribute newsletters to their members or publish upcoming events in community calendars on their websites or social media sites.

Presentations—An effective way to get families motivated to attend an upcoming Family Engineering event is to make presentations to target audiences who can promote the event to local families. Ask teachers if you can visit classrooms or a school assembly with a short activity that will get students excited about the event. Make a presentation to a school's parent/teacher organization and emphasize the positive effects of parent involvement and the quality family time that participants will experience at the event.

> **Think about publicity in the early planning stages, so you have plenty of time to create materials and distribute them in advance of your event.**

Media Coverage—If you would like a local newspaper, radio station, or TV station to attend and report on your Family Engineering event, supply them with a press release that includes all of the key information about the event. See the *Sample Press Release* on page 22.

Sample Press Release

[SCHOOL, COMPANY, OR ORGANIZATION LOGO]

For Immediate Release

Contact Information:[NAME]

                      [PHONE]

                      [EMAIL]

Families Team Up To Explore Engineering!

[NAME OF SCHOOL OR GROUP] Hosts Engineering Event For Local Families

[CITY, STATE, DATE]—Elementary-aged children and their families will gather at [LOCATION] on [DAY OF WEEK, MONTH, DAY] at [TIME] for a special [DAY, EVENING, MORNING] of fun, hands-on activities exploring the exciting world of engineering. The event is free and refreshments will be provided.

Family Engineering is a national program designed to increase public understanding and appreciation of the role engineering plays in everyday life. Family Engineering events promote creative problem solving, teamwork, and communication, as well as introduce families to the variety of exciting careers available in engineering. For additional information, or to register for the event, contact [CONTACT NAME, ORGANIZATION] at [PHONE, EMAIL, WEBSITE].

Local sponsors for this Family Engineering event include [LIST NAMES OF INDIVIDUAL OR COMPANY SPONSORS].

-END-

Your Family Engineering Event Team

For a Family Engineering event to run smoothly and provide a great experience for families, it is best to have made arrangements to have approximately one volunteer/facilitator for every 15 participants that you expect to attend.

Specific Volunteer Roles at the Event

Greeters (1-2)—Greet arriving families at a welcome table, sign them in, make or distribute pre-made nametags, and distribute raffle tickets if prizes are to be given away. It is helpful for greeters to orient the families to the event setting, and invite them to engage in the Opener activities.

Event Spokesperson—After the Opener activities, the spokesperson welcomes participants and introduces the Family Engineering event team. The spokesperson also provides closing remarks, and thanks sponsors, volunteers, and participants at the end of the event. If the event includes presentations by engineering career role models, the spokesperson can introduce the guest(s) and act as a moderator to keep track of time and assist with questions from families.

Activity Facilitators (1 for every Engineering Challenge activity)—Introduce and lead an Engineering Challenge activity. One or more facilitators may lead a number of Engineering Challenges with one large group of families, or multiple facilitators may lead different activities for smaller groups of families in separate rooms.

Activity Assistants (1 for every 15-20 participants)—Assist families as they participate in the Opener activities by encouraging participation and answering questions. Help facilitate the Engineering Challenges by assisting with materials distribution and circulating amongst families answering questions and offering encouragement. If multiple activities are occurring at the same time in different rooms, try to assign at least one activity assistant in each room.

Setup and Cleanup (3-4)—Arrive early and/or stay after the event to assist with setup and cleanup needs.

Note: It is possible that any one volunteer may end up filling multiple roles—Greeter and Activity Facilitator, for example; or all volunteers may contribute to setup and cleanup. This works as long as the roles are made clear and volunteers have the time and resources necessary to prepare for their involvement.

Recruiting Your Team

Once you have recruited a few key people to assist you with advance planning and organizing, you will want to consider other groups and/or individuals that might be available to volunteer during the Family Engineering event. Your event may be staffed completely by volunteers, or it may include individuals from a teaching staff, museum staff, or part of a community organization where the event is being held. See the next section of this chapter, "Engineers as Career Role Models," for specific information on getting engineers involved in your Family Engineering event.

> *It is best to have approximately one volunteer/facilitator for every 15 participants.*

Sources for Volunteers

* Professional engineering societies or student chapters of professional societies at local universities
* Local businesses, especially engineering and science related businesses
* Government or social services agencies
* Local parent-teacher associations (PTA)
* University schools and/or departments of engineering, technology, science, education, community outreach, equity/diversity, or admissions/recruitment

* Science centers or museums
* Youth clubs or organizations
* Scout troops
* Elementary schools
* Retired teacher associations
* Local civic groups
* High school service groups/clubs

Training Your Team

Make sure your team is prepared to deliver a fabulous Family Engineering experience by providing them with all the information and training they need to succeed. In addition to knowing the basic event schedule and responsibilities, it is important for the whole team to understand the goals of the Family Engineering program and how they can create and support an accessible and inviting learning environment for families. Give volunteers the *Working With Families: Tips for Event Volunteers* handout available in Appendix E.

Depending on the size and scope of your Family Engineering event, you may want to host a training session prior to the event day, or, simply schedule enough time on the day of the event to prepare volunteers and allow them time to become familiar with the activities. Training for event volunteers that serve as greeters, activity assistants, and/or setup and cleanup crews can be accomplished within 30-60 minutes. Activity facilitators will need additional training and/or planning time prior to the event to prepare for leading a specific activity.

Working with Families: Tips for Event Volunteers handout available in Appendix E.

> **Make sure your team is prepared to deliver a fabulous Family Engineering experience by providing them with all the information and training they need to succeed.**

General Training Session (prior to event)

Schedule a 1-1 1/2 hour training session for your event team at least a week before the event. It is helpful if you can do the training in the same location as the event, but it's not necessary. Have the Opener activities set up on tables, or, better yet, have the volunteers set them up, so that everyone can try them out. Bring examples of the Engineering Challenge activities that will be done at the event, and, if time permits, have volunteers engage in one or more of these activities.

Sample General Training Session Agenda

♦ Allow time for exploring the Opener activities (15-20 minutes)

♦ Introduce the Family Engineering program philosophy and goals (5 minutes)

♦ Review event schedule and volunteer roles (15 minutes)

♦ Discuss effective strategies for working with families and distribute *Working With Families: Tips for Event Volunteers* handout (Appendix E) (5 minutes)

♦ **Openers**—what they are, how to set up and clean up, how to monitor activity areas, tips for assisting families (10 minutes)

♦ **Engineering Challenges**—what activities will be done (show examples or try them out), how to distribute materials, tips for assisting families, etc. (10-30 minutes)

Volunteer Orientation (day of event)

Schedule a volunteer orientation on the day of your event, preferably after the event has been set up and is ready to go. Use this time to review the event schedule, orient the team to the facility, assign or confirm volunteer roles, and distribute nametags. If the full team has already attended a general training session, this orientation can be brief (20-30 minutes). However, if this is the only training your event team will have, schedule at least an hour in order to also cover the material listed above in the sample general training session agenda. Complete the volunteer orientation at least 1/2 hour before the event start time so that your team can take care of last minute details and be prepared for any families who arrive early.

Activity Facilitator Training and Preparation

Anyone who will lead an Engineering Challenge at the event should prepare in advance by reading the activity thoroughly, trying it out themselves, and planning for how to facilitate the activity with families. Activity facilitators should attend a general training session and then work with you individually to prepare for their specific activities. Or, if you have a number of facilitators, you can work with them as a group and have them participate in an activity while you model effective facilitation strategies. For convenience, training the activity facilitators as a group can take place immediately following the general training session for the full team. Be sure to introduce them to the engineering design process information found in "The World of Engineering" chapter, and share the following tips with the activity facilitators for your event.

Tips for Engineering Challenge Activity Facilitators

A Family Engineering event is an informal learning experience with a focus on increasing interest and awareness of engineering. The focus of the activities is not on teaching or learning specific engineering concepts or content. Rather, the activities are designed to use engineering content to create positive experiences with engineering and to encourage family interaction. As an activity facilitator, you play a key role in supporting this approach and encouraging fun, hands-on exploration of engineering.

Be Prepared.
- Try the activity yourself in advance.
- Have materials gathered and prepared for easy distribution.

Create a relaxed and welcoming atmosphere by having fun and enjoying your interactions with the families.

Focus on families working together.
- Make sure that families sit together as a family. Two small families can work together on an activity if they wish or if you have a limited number of supplies.
- Limit the amount of time you talk or present to the group so that families have plenty of time to work together on the hands-on Engineering Challenge.
- Encourage families to find solutions and explore different design options, rather than giving them answers or showing them how to accomplish a challenge.
- Encourage families to improve their own design and find success by accomplishing a challenge, rather than competing with other families.
- Give families a chance to share their work with each other and discuss their successes and challenges.

Make connections to engineering.
- Use real-life engineering examples that relate to the activity and to the families attending.
- Point out when families are using engineering skills such as problem solving, teamwork, and communication.
- Review the information on the engineering design process found in "The World of Engineering" chapter and remember to emphasize the process over the outcome, pointing out when families are engaged in a step of the design process.
- Emphasize that there can be many different solutions to a problem and remind families that engineers often attempt many designs before settling on a final one.

Engineers as Career Role Models

A Family Engineering event is an excellent opportunity to introduce children and their parents to the variety of engineering fields and careers. A career role model is someone who is interested in sharing the excitement of his/her work with others. This may be a professional engineer in your community or an engineering student from a local college or university. Meeting career role models and having positive experiences with engineers are ways for children to discover that engineering is a rewarding and realistic career option for them. It may even be their first time meeting an engineer! A Family Engineering event can be more meaningful and effective when career role models are involved as volunteers, facilitators, or guest speakers.

Recruiting Engineers as Volunteers and Career Role Models

You may be an engineer yourself or already have engineers or engineering students involved on your event planning team. If not, you can recruit career role models from local engineering firms, engineering societies, universities, city or state agencies (transportation and city planning, utilities, recycling, or waste management departments), and/or local industries (health care, pharmaceuticals, construction, technology, manufacturing). You may even be able to find an engineer among the parents or families that will attend the event. It can be a powerful experience for children and parents to meet someone within their own community that is trained and working as an engineer.

The diversity of the workforce should be reflected in the role models you recruit for the event, with special consideration for representing the diversity of your target audience. Make an effort to invite women, people from diverse ethnic and racial backgrounds, or physically challenged individuals that work in engineering fields. Also, be aware of representing a variety of fields of engineering whenever possible, so that families see the diversity of opportunities and interests that can be pursued within engineering.

Key Information for Recruiting Engineers

* Why is it important? (This is an opportunity for them to inspire children and parents, raise public awareness, and increase understanding of engineering and engineering careers.)

* What is a Family Engineering event like? (an informal, relaxed, and friendly environment for the whole family)

* Who will attend the event? (age of children, number of participants, etc.)

* What will be their role at the event? (speaker, volunteer, activity facilitator, etc.)

* Will they be the only career role model at the event?

* What is the time commitment? (including orientation, advance preparation, or clean up)

* Who should they contact with any questions?

Engineers at your event will be most successful if they are comfortable interacting positively with families and can communicate effectively about their work in a fun and engaging way. When possible, seek out career role models that already have the experience and skills necessary to work with families. Be sure to provide volunteers with the guidance and information that will lead to a successful Family Engineering experience for everyone.

Preparing Engineers as Volunteers and Career Role Models

Engineers can play a number of roles at a Family Engineering event, including helping families during Opener activities, handing out materials/activity supplies, facilitating an Engineering Challenge activity, or serving as a guest speaker. If time allows, ask engineers to speak briefly (4-5 minutes each)

> *Engineers can play a number of roles at a Family Engineering event.*

about their work and answer questions from the families. You may want to have someone act as a moderator to assist with the question and answer session and to keep track of time so that sufficient time remains for the rest of your event activities. No matter how they are involved, be sure to introduce the engineers and their engineering fields early on in the event so that families know who they are. Encourage engineers to interact with families throughout the event.

To make this a successful experience for everyone, prepare engineers for the type of information that may interest families, as well as for the types of questions engineers may be asked by giving them the *Working With Families: Tips for Engineers* handout available in Appendix E. Prepare engineers for general event volunteering or facilitating an activity by providing them with the orientation and training described in the previous section—"Your Family Engineering Event Team."

> *Working with Families: Tips for Engineers* handout available in Appendix E.

Advance Preparation

Planning ahead, spreading the work out over time, and preparing materials and supplies in advance will make the week of the event more manageable and the event itself more enjoyable.

Getting Organized

☐ Use the *Family Engineering Event Planning Checklist* (Appendix E) to guide your preparation.

☐ Review the next section—"At the Event"—to help you visualize what your event will look like and what supplies and resources you will need.

☐ Create your own event schedule using the *Sample Event Schedules* on pages 30-31.

Activity and Materials Preparation

☐ Read each activity description thoroughly and try it out yourself.

☐ Make sure that other activity facilitators have also tried out their activities in advance.

☐ Review the "Supplies" section for each planned activity and purchase or gather supplies.

☐ Review the "Advanced Preparation" section for each planned activity to determine what preparation steps can be done before the day of the event.

☐ Prepare supplies and copy materials for Openers and Engineering Challenges.

Openers—For the Opener activities, it is helpful to gather all the materials for each Opener, including signs and/or consumables, into a separate extra-large, resealable plastic bag that is labeled with the Opener activity's name. This makes setup and cleanup easy and the supplies are already organized for your next event!

Opener activity signs and activity interactives are located in the Appendix. The activity signs should be two-sided and copied in color onto white cardstock whenever possible. It is most effective to use a document clip or holder that allows the sign to stand upright next to the activity. The interactive materials that are used in several of the Openers should also be copied in color on white cardstock. If you anticipate using these signs or interactive materials for more than one event, it is best to laminate them.

Engineering Challenges—The supply list for each Engineering Challenge activity tells you what supplies are needed for **one family or small group**. When buying or gathering supplies, be sure to multiply this list by the number of families you expect to attend your event and then have a few additional supplies on hand just in case. Many of the Engineering Challenges recommend that you create a supply bag for each participating family in order to speed up materials distribution at the event. To do this, it is helpful to gather a few volunteers, lay out the supplies, and use an assembly line approach for filling the supply bags. If you choose not to create family supply bags, you can also lay out the supplies on a table at the event and invite families to collect them or have volunteers hand them out separately. *Design Challenges*, the handout pages for the Engineering Challenges, are available in Appendix C and can be copied in black and white on regular white or colored copy paper. Some Engineering Challenges also have interactive materials that are available in Appendix D.

Sample Event Schedule for Single Large Group Event Format
6:00 - 8:00 pm

4:00 - 5:15 pm — **Event Setup**
- ☐ *Setup crew and activity facilitators arrive*
- ☐ *Finalize table arrangements*
- ☐ *Test sound system (if using one)*
- ☐ *Set up Opener activities*
- ☐ *Set up supply tables for Engineering Challenge activities*
- ☐ *Set up welcome table*
- ☐ *Set up refreshments*
- ☐ *Post welcome and directional signs*

5:15 - 5:45 pm — **Event Volunteer Orientation**
- ☐ *Additional volunteers arrive*
- ☐ *Review event schedule and assign or confirm responsibilities*
- ☐ *Distribute volunteer nametags*

5:45 - 6:15 pm — **Families Arrive, Opener Activities Available**
- ☐ *Sign in and make or distribute nametags*
- ☐ *Families interact with Opener activities*
- ☐ *Refreshments available*

6:15 - 6:30 pm — **Families continue to interact with Opener activities**

6:30 - 6:40 pm — **Welcome and Introductions**

6:40 - 6:50 pm — **Engineer/Career Role Model Presentations (optional)**

6:50 - 7:50 pm — **Engineering Challenge Activities**
One activity facilitator leads 1-3 activities for all participants at same time

7:50 - 8:00 pm — **Wrap Up**
- ☐ *Ask families to complete evaluation forms*
- ☐ *Thank sponsors, families, and volunteers*
- ☐ *Encourage continued engineering exploration at home*
- ☐ *Distribute Exploring Engineering at Home handout*
- ☐ *Raffle prizes (optional)*

8:00 - 8:30 pm — **Event Cleanup**

FAMILY engineering

Sample Event Schedule for Multiple Small Groups Event Format
6:00 - 8:00 pm

4:00 - 5:15 pm **Event Setup**
- ☐ *Set up crew and activity facilitators arrive*
- ☐ *Finalize table arrangement in all rooms*
- ☐ *Test sound system (if using one in large gathering area)*
- ☐ *Set up Opener activities*
- ☐ *Activity Facilitators set up individual rooms and supplies for Engineering Challenge activities*
- ☐ *Set up welcome table*
- ☐ *Set up refreshments*
- ☐ *Post welcome, directional, and room signs*

5:15 - 5:45 pm **Event Volunteer Orientation**
- ☐ *Additional volunteers arrive*
- ☐ *Review event schedule and assign or confirm responsibilities*
- ☐ *Distribute volunteer nametags*

5:45 - 6:20 pm **Families Arrive, Opener Activities Available**
- ☐ *Sign in and make or distribute nametags*
- ☐ *Families are assigned or sign up for Engineering Challenge activities*
- ☐ *Families interact with Opener activities*
- ☐ *Refreshments available*

6:20 - 6:30 pm **Welcome and Introductions**

Engineer/Career Role Model Presentations (optional)

6:30 - 7:50 pm **Engineering Challenge Activities**
Multiple activity facilitators lead 30-40 minute activities for small groups of 20-30 participants in separate rooms. After 30-40 minutes, families rotate to a second activity room.

| 6:30– 7:10 pm | Engineering Challenge Session One |
| 7:15 -7:55 pm | Engineering Challenge Session Two |

7:55 - 8:00 pm **Wrap Up (completed by activity facilitator in final activity room)**
- ☐ *Ask families to complete evaluation forms*
- ☐ *Thank sponsors, families, and volunteers*
- ☐ *Encourage continued engineering exploration at home*
- ☐ *Distribute Exploring Engineering at Home handout*
- ☐ *Raffle prizes (optional)*

8:00 - 8:30 pm **Event Cleanup**

At the Event

The schedule that you choose for your event can vary. Most successful events are between 1 ½ to 2 hours long. If a meal is being served, then you may want to extend this time by half an hour. An event that is much longer than this may require too great a commitment on the part of the families or your volunteers. This section will take you step-by-step through a Family Engineering event. The *Sample Event Schedules*, found on pages 30-31, provide you with an example for how to organize your event day.

> **Allow plenty of time for setting up the event.**

Event Setup

Allow plenty of time for setting up the event so that you are ready 15-30 minutes before the scheduled event time and can welcome early arrivals. Schedule a few volunteers to assist with setup and all others to arrive in time for a volunteer orientation prior to the start of the event.

Signs and Posters—Make welcome posters and directional signs to greet families and direct them to the proper rooms for the event. Place signs on the outside of the facility and even in the parking lot to help families find their way to the event entrance. Post signs inside that will direct families to restrooms and activity areas. If multiple activities are scheduled in separate rooms, post event schedules with room locations, and signs with the activity names next to the appropriate rooms.

Welcome Table—Place a table at the event entrance where families can check in, make or distribute pre-made nametags, and receive an event program (if desired). A sample *Family Engineering Event Sign-In Sheet* is available in Appendix E. If you are planning a raffle for prizes later in the event, families can receive a raffle ticket or write their family name on a slip of paper to place in a jar for the prize drawing.

Nametags—Using nametags is a good idea at a Family Engineering event. They help you identify and call people by name and they help participants get to know each other. First names, in large letters, work the best! If making these on-site, provide blank nametag stickers and colorful, dark colored markers. Don't forget to place a trash receptacle nearby for the peeled backs of nametags.

Refreshments—Everyone loves snacks! It is nice to have simple, healthy refreshments available for participants. Keep food tables separate from the activity tables and make sure there are trash receptacles nearby. Refreshments can be as easy as apple slices or grapes and graham crackers. Snacks can be available throughout the event or scheduled for a specific time. If you are serving a more substantial meal as part of the event, it is best to have the meal scheduled before or after the activities, or in another room.

Microphone and Sound System—If the room is larger than a typical classroom, having a microphone and sound system will greatly facilitate everyone being able to hear instructions.

Room Layout—Your Family Engineering event will need space for families to engage in both the short, self-directed Opener activities that are set up on tabletops, and the longer, facilitated Engineering Challenge activities that are done while seated at tables. It is best to have the Openers set up on dedicated tables so that they do not need to be cleared away before starting the Engineering Challenges. Multiple Opener stations can be laid out on a single table, but you will want to make sure there is enough space for families to gather around the materials and work together, as well as adequate space between tables to maintain smooth traffic flow. The diagram below depicts a basic room layout for an event where all activities are occurring in one room. Although your event format or available space may be different, the basic elements are the same.

ROOM LAYOUT

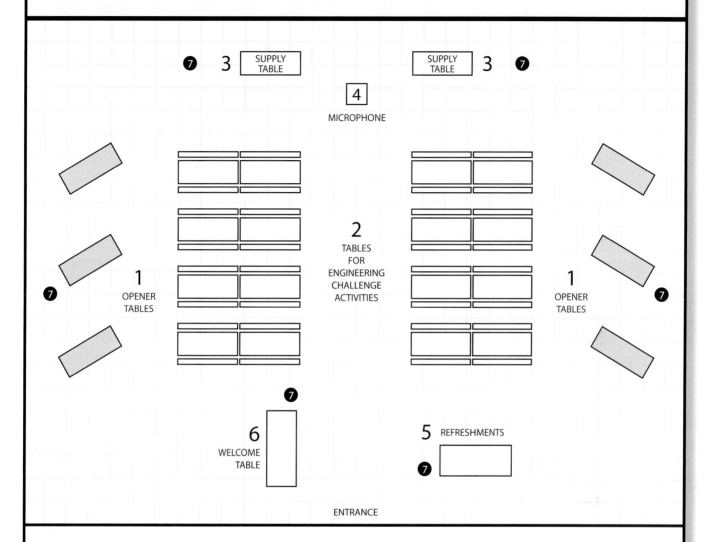

1. Opener tables placed around the perimeter of the room

2. Engineering Challenge tables and chairs that will accommodate the full group with 1-2 families seated at each table

3. Supply tables placed at the front or side of room for Engineering Challenge materials

4. Microphone and sound system at the front of the room

5. Refreshment table located away from the activities

6. Welcome table at the entrance of the room or building

7. Trash receptacles placed at the appropriate Opener tables, refreshment table, welcome table, and in Engineering Challenge activity areas as needed

Opener Activity Setup—Set up each Opener activity with the appropriate materials and activity sign. If you have packed each Opener into a separate, extra-large resealable plastic bag, it will be easy for volunteers to quickly take these bags to the tables and lay out the materials. Be sure to keep the bags handy for activity materials to be returned during cleanup. Opener activities that have consumable materials should have a trash receptacle nearby.

Engineering Challenge Activity Setup—The materials for each Engineering Challenge activity should be placed on the supply tables near where the activity will be facilitated. Materials that will be gathered by families during the activity should be arranged on supply tables that are easily accessible to participants. If more than one activity will be conducted in the same area, have each set of materials ready to go and prepare volunteers to assist the facilitator with managing the transition between activities.

Some of the Engineering Challenge activities call for a "testing area" where families can test out their built designs. The size of this area will depend on the specific activity and your available space. It is helpful to post a sign so that families can easily locate the "testing area." This can be located in the back or front of a room, in a nearby hallway, or in a separate room if necessary.

Families Arrive!

Be prepared for families that may arrive early by being ready at least 15-30 minutes before the scheduled event start time. The welcome table provides a good opportunity for volunteers to greet families and orient them to the event setting. After families check in and obtain their nametags, they should be encouraged to engage in the Opener activities as a family. It helps to let families know that they do not need to do these activities in any particular order or complete every one of them, and that they can move through them according to their own interest and pace. The Opener activities allow for both early and late arrivals to feel comfortable and able to jump right into the action. During this time, volunteers should circulate among the families to encourage participation, answer questions, and check on the activity materials.

Welcome and Introductions

After allowing 20-30 minutes for families to engage with the Opener activities, invite everyone to sit together as families. If appropriate, have someone from the community, sponsor organization, or host facility do some portion of the welcome message. Welcome the families to the event and stress that this event is about families learning together. Provide a brief overview of the event and any general information about the facility, such as restroom locations, refreshments, etc. Introduce the volunteers in the room by having them all stand or come up to the front. Introduce any special guest engineers or engineering students in attendance. If you are including an engineering career role model presentation, this may be a good time to do it because you have the full group's attention prior to starting the Engineering Challenge activities. Another alternative is to schedule this brief presentation between two of the longer activities.

...remind families to have fun exploring engineering!

Most important—remind families to have fun exploring engineering!

Tips for a Smooth Event

Part of having a successful event is managing the needs of a large group of people effectively. For example, having Opener activities available at the beginning of the event allows both early and late arrivals to become immediately engaged in the event and eliminates the "hurry up and wait" feeling while everyone is arriving. Also, having plenty of tables and spreading them out for easy traffic flow creates a comfortable space with room for everyone. Think through your event schedule and plan ahead for times when families may be moving to a new location, transitioning from one activity to another, or needing instructions about what is next. The facilitated Engineering Challenge activities may be done with one large group or with smaller groups in separate rooms. Here are some tips for managing these different formats.

> ***Single Large Group***—When using this format, it is important to plan ahead for the transitions between activities to minimize the amount of time the families spend waiting for the next activity to begin. Have the materials for each activity staged and ready for quick distribution. When an activity is finished, ask families to repackage supplies and hand them to a volunteer or return them to a supply table. Also ask them to dispose of any trash.

> ***Multiple Smaller Groups***—When using this event format, you will need to recruit and train several activity facilitators (one per 20-25 participants). Providing one activity assistant per room is also recommended. It is important to remind facilitators that they need to complete their activities in the allotted times. If they finish earlier than expected, they should be prepared to conduct an Icebreaker activity (see "Icebreaker Activities" on page 37) that can be used to keep families engaged while waiting to move on to their next activity. It is also desirable to spread families out evenly among the different activity rooms. One way to assure that participants are evenly distributed is to have a sign up sheet at the welcome table and small 'reservation' cards to help families remember which activities they are attending at which time.

If you plan to hold multiple Family Engineering events, you will want to save any items that can be reused, as well as any extra consumable materials. Ask families to return unused, clean materials to the bags or containers so that you can restock the same bags for the next event. You will want to keep used bags separate from unused bags so you know which ones will need to be restocked.

Evaluation

We all benefit from feedback. Asking participants for their feedback at the end of your Family Engineering event will let you know how the event went and can help you improve future events. If the event was a success, sponsors and supporters will enjoy hearing specific details. Documenting the success through participant feedback will help you repeat that success. It may also help to secure future support.

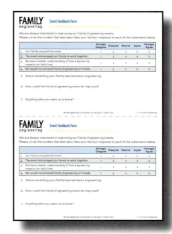

A sample *Family Engineering Event Feedback Form* is provided in Appendix E. You can simply copy and distribute this form to families, or you can make one of your own. Remember to keep it short and simple for the best response. Be sure to give families plenty of time to fill it out. If they are rushed, their answers will not be as thorough or as thoughtful as you want. Encourage the entire family to complete the form together. Also, it is helpful to have families complete the feedback form prior to any closing remarks.

Sample *Family Engineering Event Feedback Form* available in Appendix E.

> ❗ *Taking the time to ask participants for their feedback at the end of your Family Engineering event will let you know how the event went and can help you improve future events.*

Closure

At the end of your event, take the time to communicate some general messages about engineering, such as the fact that engineers are creative problem-solvers or that engineering and engineered products are a part of our daily lives. Distribute the *Exploring Engineering at Home: Tips for Parents and Other Adult Caregivers* handout (in Appendix E) and encourage families to continue exploring engineering at home and in their community. Thank event sponsors and all the families and volunteers for attending. If you are including raffle prizes as part of your event, you should hold the raffle at the very end, following the collection of the feedback forms and any closing remarks, to discourage families from leaving the event early.

Exploring Engineering at Home: Tips for Parents and Other Adult Caregivers handout available in Appendix E.

Icebreaker Activities

Icebreakers are fun group activities that can be done in 5-10 minutes with no materials. These quick activities can be used to kick off an event or during a transition time to get families relaxed and everyone's minds thinking about engineering.

I Spy Engineering

Explain to families that engineers design and create many things to solve all kinds of problems, and that there are examples of engineering all around us. Ask them if they see anything in the room that was created to solve a problem. Invite families to play "I Spy Engineering" as part of a large group or with their family. One family member "spies" an example of engineering and others try to guess what it might be. Introduce and model the game with a few examples:

> "I spy with my little eye something an engineer designed so that we would not have to sit on the floor." (chair)

> "I spy with my little eye something an engineer designed to solve the problem of dividing one room from another." (walls)

Are You an Engineer?

Ask families for a show of hands if they have ever done any engineering. Read the following statements out loud, asking families to raise their hands when a statement describes something they have done. Then remind families that engineering isn't something that only other people do! We all do different kinds of engineering everyday!

1. Who has created structures with blocks? (civil engineering)

2. Who has used rocks or other objects to change the direction of water flow in a stream? (civil engineering)

3. Who has designed a system to organize household paperwork? (systems engineering)

4. Who has designed and built a gingerbread house? (civil engineering, structural engineering)

5. Who has taken something apart and tried to make it better? (mechanical engineering)

6. Who has designed their own food "concoction" or recipe? (chemical engineering)

7. Who has kept the water clean in a tank to keep a frog or fish alive? (environmental engineering)

8. Who has built a little boat for floating in a bathtub, lake, or stream? (mechanical engineering)

9. Who has created a route for a bike ride or car trip? (transportation engineering)

10. Who has figured out how to fix a computer hardware problem? (computer engineering)

11. Who has designed a paper airplane? (aerospace engineering)

Agree or Disagree?

Tell families that you are going to read some statements about engineering and they are going to decide whether they agree or disagree with what you say. Designate one end of the room as "Agree" and the other as "Disagree." Decisions can be shown by physically moving to one end of the room, by pointing to that end of the room or, if undecided, staying in the middle or pointing straight up. Read each of the following statements about engineering. After giving time for families to make a decision, read the suggested explanation in italics after each statement.

"Most engineers build bridges and roads." *(DISAGREE) People often think of engineers and the roads and bridges they design, which are very valuable. However, in reality, this is just one area of engineering (civil), and other types of engineers work on many different types of problems, including cleaning our drinking water, designing new artificial body parts, designing buildings, designing fuel efficient automobiles, etc.*

"There are jobs for engineers all around the world." *(AGREE) Engineers do work in all parts of the world and they represent all ethnicities and cultures.*

"Men make better engineers than women." *(DISAGREE) Both men and women make excellent engineers. Currently more men become engineers, that is one reason we're here: to let girls know that they can be engineers, too!*

"Anyone can become an engineer." *(AGREE) Curious kids who enjoy learning how things work, like to design things to help other people, and like math and science, will probably enjoy becoming engineers. To be an engineer, you need to go to college after graduating from high school.*

"Engineers have all the answers." *(DISAGREE) Engineers are always working to learn more and improve products and ways of doing things, but no one has "all the answers."*

"Engineers always work with machines." *(DISAGREE) Computers are the only machines that most engineers use on a daily basis. Only a few types of engineers use other machines daily.*

"Engineers usually do their work alone." *(DISAGREE) Most engineers work with other people in teams, just as in many other types of jobs.*

"Some engineers develop new technologies to help sick or injured people feel better." *(AGREE) Many people don't realize that engineers help design new drugs to cure disease, as well as medical materials and devices to help people walk, see, or hear again. This field of engineering is called biomedical engineering.*

"Engineers help develop solutions for everyday problems and challenges." *(AGREE) The products we buy to help us solve problems and meet the needs of everyday life, such as can openers, ipods, cars, computers, and cleaning products, are designed by engineers.*

"Engineers help develop solutions for the world's big problems. They help address issues such as keeping water clean, improving human health, developing new energy sources, and growing food for a rapidly growing population." *(AGREE) Many engineers are involved in helping to develop solutions to some of the world's biggest problems.*

Add additional statements, as appropriate.

OPENERS

6

The activities in this section are called "Openers" because they open the doors to learning through short, fun experiences with engineering. Openers are designed to be quick, hands-on activities that encourage families to experiment, tinker, and solve problems together. They are self-directed, tabletop activities that families can engage in at their own pace. When used at the beginning of a Family Engineering event, they help to get participants comfortable in the event setting, and actively engaged in engineering tasks and topics soon after their arrival at the event.

Openers typically require a tabletop or flat surface, an activity sign, and some simple, inexpensive materials. The supplies are listed at the beginning of each Opener activity, and activity signs for each activity are available in Appendix A. The two-sided activity signs have instructions or prompts for the activity on the front and a description of the activity's real-life connection to engineering on the back.

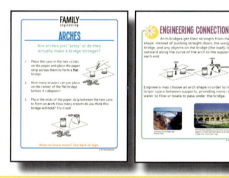

Opener activity signs are available in Appendix A.

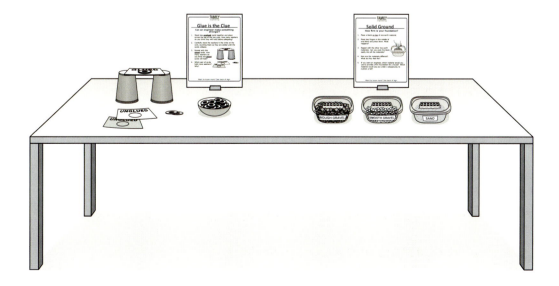

Opener Activity	Page Number	Difficulty Level	Activities That Work Best with Multiple Stations	Engineering Fields
Against the Wind	43	●		Mechanical, Aerospace
All the Right Tools	45	○		General
Arches	46	○	√	Civil
Boxing Beans	48	○	√	Package, Materials
Domino Diving Board	50	◑	√	Civil, Mechanical
Glue is the Clue	52	◑	√	Materials, Chemical
Happy Feet	54	○		Biomedical, Materials, Industrial
Inspired by Nature	56	○		General
Learning From Failure	58	◑	√	General
Let's Communicate	60	◑	√	General
Make It Loud!	62	○		Materials, Biomedical, Mechanical
Picture This	64	◑		General
Shifting Shapes	66	◑		Civil
Showerhead Showdown	68	○		Environmental, Mechanical
Solid Ground	70	○		Geological, Civil
Soundproof Package	72	○	√	Acoustical, Materials, Package
Thrill Seekers	74	◑	√	Mechanical
Tumbling Tower	76	○	√	Civil
What Do Engineers Do?	77	○		General
Who Engineered It?	78	●		General
Wrap It Up!	80	◑	√	Materials
Your Foot, My Foot	82	○		General

Difficulty Level

○ Simplest

◑ Moderate difficulty

● More complex challenges

Opener activities are designed so that they can be completed in a short amount of time (3-5 minutes), allowing families to experience a variety of these activities at the opening of an event. However, a family may also become engaged in one activity for a longer period of time if a topic or a challenge holds their interest. For this reason, it is often helpful to have multiple stations of the same activity available so that several families can work on a given activity at the same time. Opener activities that benefit most from multiple stations are noted in the "Openers At-A-Glance" chart.

Think about the available time and space for doing the activities, and the interests and needs of the families attending the event.

When families are doing Opener activities, event volunteers can be available to answer questions, assist families, or offer encouragement. However, volunteers should avoid demonstrating or telling families how to solve a particular challenge. Instead, volunteers should encourage parents to work with their children to read the instructions and try the activity as a family. Make sure there is enough space around each Opener for a family to gather around the materials together, and that activity tables are spaced far enough apart to allow for a comfortable flow of traffic between Opener activities.

Selecting Opener Activities

When choosing Opener activities to include in a Family Engineering event, think about the time and space available for doing the activities, the interests and needs of the families attending the event, and whether the chosen set of activities represents a diversity of engineering fields and topics.

The recommended number of different Opener activities or duplicate stations that should be offered depends on the expected number of participants, as well as the amount of time allowed for exploring the Openers. If there is only a short amount of time scheduled, setting out too many different activities may make families feel pressured to complete them all, or make them feel as if they are missing out if they are not able to participate in all the activities within the allotted time. On the other hand, if there are not enough different activities available, families may run out of things to do. Adding more activities or duplicate stations of any one activity can help to accommodate a higher number of participants or a longer scheduled time for Openers. In general, it works best to have approximately one station available for every 4-5 participants.

Recommended Number of Opener Activities for Various Family Engineering Events

Number of participants	Scheduled Time for Opener Activities		
	15 Minutes	30 Minutes	40 Minutes
25-50	4-5 Activities 1-2 Stations for each	8-9 Activities 1-2 Stations for each	10-12 Activities 1-2 Stations for each
50-75	4-5 Activities 2-3 Stations for each	8-9 Activities 1-2 Stations for each	10-12 Activities 1-2 Stations for each
75-100	5-6 Activities 2-3 Stations for each	9-10 Activities 2-3 Stations for each	11-12 Activities 1-2 Stations for each
100-150	5-6 Activities 3-4 Stations for each	9-10 Activities 2-3 Stations for each	12-13 Activities 1-3 Stations for each

Introducing families to the many different fields and career paths in engineering, as well as broadening understanding of what engineers do, are two of the most important goals of Family Engineering. Make an effort to choose activities that represent different engineering fields and/or highlight different aspects of engineering. Use the "Openers At-A-Glance" chart to help you choose activities that meet your event needs, but also give families a broad exposure to engineering. Remember to consider which Engineering Challenge activities families will experience during the event, and select Openers that complement these challenges, or ones that introduce different engineering concepts and skills for variety.

Sample Opener Activity Selections

Six Opener Activities

Glue is the Clue		Domino Diving Board
Learning From Failure	*or*	Happy Feet
Make It Loud!		Inspired by Nature
Showerhead Showdown		Solid Ground
Thrill Seekers		Tumbling Tower
Who Engineered It?		Wrap It Up!

Eight Opener Activities

Arches		Against the Wind
Boxing Beans		Glue is the Clue
Domino Diving Board	*or*	Inspired by Nature
Happy Feet		Let's Communicate
Shifting Shapes		Make It Loud!
Showerhead Showdown		Solid Ground
Soundproof Package		Tumbling Tower
Who Engineered It?		Thrill Seekers

Twelve Opener Activities

All the Right Tools		Boxing Beans
Arches		Domino Diving Board
Against the Wind	*or*	Glue is the Clue
Inspired by Nature		Happy Feet
Learning From Failure		Let's Communicate
Make It Loud!		Shifting Shapes
Picture This		Showerhead Showdown
Shifting Shapes		Solid Ground
Soundproof Package		Thrill Seekers
Thrill Seekers		Tumbling Tower
What Do Engineers Do?		Who Engineered It?
Wrap It Up!		Your Foot, My Foot

How can engineers save energy through design?

Advance Preparation

- Cut index cards in half so that they are approximately 2½" x 3".

- Attach a card piece to each car in a different design using rubber bands and tape (see illustration below). Make sure that the wheels are free so that they still roll smoothly. Have at least one car that will not be affected by the wind when placed in front of the fan, and at least one or two cars that will catch the wind and slow down.

- Place one end of the cardboard piece on the edge of the book to create a ramp and place the fan about 18-20" from the bottom edge of the ramp.

- Test out all the cars to make sure the fan will provide enough wind to slow down a few of the cars.

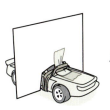   

Engineering Fields

- *mechanical engineering*
- *aerospace engineering*

Engineering Concepts & Skills

- *optimization/tradeoffs*
- *sustainability*
- *modeling*
- *controlled experimentation and testing*

Supplies

- *3-5 identical toy cars of the same size that roll smoothly on a tabletop (Matchbox® cars work well)*
- *2-3 small index cards (3" x 5")*
- *12" x 20" piece of stiff, smooth cardboard*
- *large book, about 1-1½" thick*
- *small clip-on or box fan*
- *Against the Wind activity sign (Appendix A)*

Advance Preparation Supplies

- *scissors*
- *rubberbands*
- *tape*

ENGINEERING CONNECTION

At highway speeds, most of the energy (fuel) needed to keep a car moving down the road is used to push air out of the way. Engineers can help us save energy by designing more **aerodynamic** cars and trucks. This means they have minimal air resistance and move through the air easily. Some strategies for making a car more aerodynamic include changing the shape of the car, making rearview mirrors smaller or placing them inside the car, covering the wheel openings, and lowering the car so that it is closer to the ground.

Activity Steps

1. Turn on the fan and hold one car at the top of the ramp. Let go of the car, allowing it to roll straight toward the fan. What happens?

2. Try again with another car until you have tested all the cars. What do you notice?

3. Which car is the most aerodynamic (moves easily through the wind)? Which car do you think would need to use the most energy (fuel) to move against the wind?

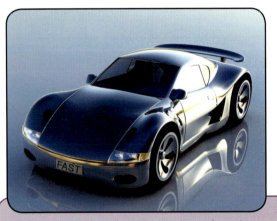

This car is designed to be aerodynamic and will experience minimal air resistance as it moves through the air.

This car **is not** designed to be aerodynamic and will experience a lot of air resistance as it moves through the air.

ALL THE RIGHT TOOLS

Can you choose the best tool for a job?

Engineering Fields
- *general engineering*

Engineering Concepts & Skills
- *communication*

Supplies
- *ruler*
- *yardstick*
- *tape measure*
- *½ cup measuring cup*
- *measuring spoons*
- *stopwatch or timer*
- *small bowl containing at least ¾ cup of rice*
- *All the Right Tools activity sign (Appendix A)*

Activity Steps

Engineers rely on accurate measurements to do their work. Which measuring tool will work best to accurately measure each of the items below?

- Width of the table
- Length of your finger
- Small amount of rice
- How long it takes to walk around the table
- Width of your thumbnail
- Large amount of rice
- Length of the room
- Length of your arm from elbow to wrist
- Your height

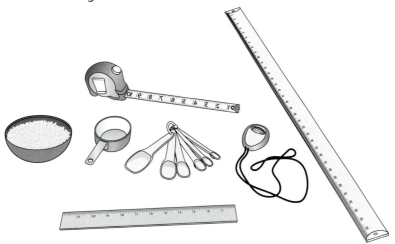

⚙ ENGINEERING CONNECTION

Taking accurate measurements is very important in engineering. Engineers use accurate measurements to help them draw a design, construct models, or give instructions to the people who will create or build a product the engineer has designed.

ARCHES

Engineering Fields

- *civil engineering*

Engineering Concepts & Skills

- *controlled experimentation and testing*
- *modeling*

Supplies

- *4" x 12" strip of heavyweight poster board*
- *2 unopened cans of food of the same size*
- *8 identical rectangular erasers to use as weights*
- *sheet of paper (11" x 17")*
- *Arches activity sign (Appendix A)*

Advance Preparation Supplies

- *clear tape*
- *marker*

Are arches just "artsy" or do they actually make a bridge stronger?

Advance Preparation

- On the sheet of paper, use the marker to trace the base of a can, making two circles that are 6" apart.

- Tape the paper mat to the table and arrange materials as in the illustration below.

- *Event Tip:* You may want to have 2-3 extra strips of poster board available to replace the original strip if it becomes bent or creased.

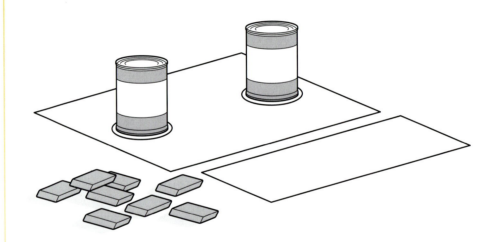

⁂ ENGINEERING CONNECTION

Arch bridges get their strength from their shape. Instead of pushing straight down, the weight of the bridge and any objects on the bridge (the load) are carried outward along the curve of the arch to the supports at each end. Engineers may choose an arch shape in order to have a larger space between supports, providing more room for water to flow or boats to pass beneath the bridge. Although arch bridges were made of stone in the past, today materials like concrete or steel allow engineers to design longer arches spanning greater distances.

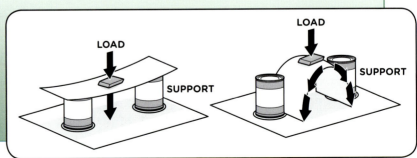

ANCIENT ARCHES

Arch bridges are naturally strong and were one of the earliest types of bridges constructed. Over two thousand years ago, the Romans built a large aqueduct (a structure to supply water) above an arched bridge in France. Built on three levels, the Pont du Gard has a total of 64 arches! The bridge was made with stones cut and placed so carefully that mortar was used only in its top tier. The stones in the two lower tiers stay together by the force of the weight above. It is still possible to cross the Pont du Gard on the lower level stone road.

Activity Steps

1. Place the cans in the two circles on the paper. Then place the paper strip across the cans to form a flat bridge.

2. How many erasers can you place on the center of the flat bridge before it collapses?

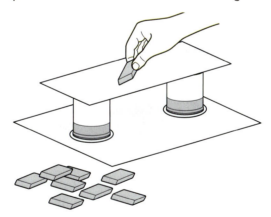

3. Place the ends of the paper strip between the two cans to form an arch. How many erasers do you think this bridge will hold? Try it out!

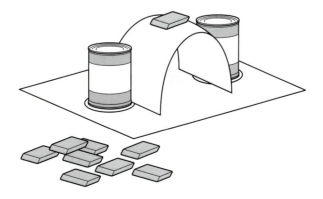

BOXING BEANS

Supplies

- *one cup of dried beans in a bowl*
- *4 pre-assembled 3-dimensional shapes made of heavy cardstock (Shape Templates in Appendix B)*
- *small paper cup*
- *cookie sheet or plastic tray (optional)*
- *Boxing Beans activity sign (Appendix A)*

Advance Preparation Supplies

- *scissors*
- *tape*

Are packages engineered?

Advance Preparation

- Copy the four different *Shape Templates* onto heavy cardstock.

- Use scissors and tape to cut out and assemble each 3-dimensional shape. Leave one side of each shape open (un-taped) for filling with beans.

- Place beans and small paper cup, for scooping, into a bowl.

- Place materials on the table as shown in the illustration below.

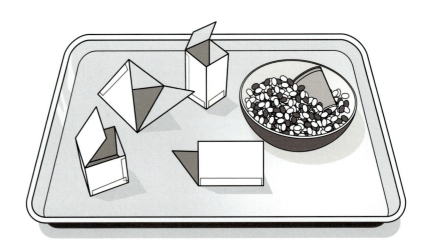

Does packaging influence what you buy?

Activity Steps

1. Predict which package shape will hold the most beans.
2. Fill one package to the top with beans so that the lid can still close.
3. Pour the beans directly into another package. Does this new package hold more or less?
4. Repeat with all of the packages. What did you discover?
5. If you were an engineer, which package shape would you use to:
   - Make it appear that it holds the most product?
   - Stack easily on a shelf?
   - Attract attention with a unique shape?

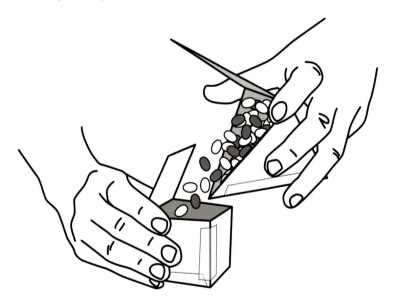

✴ ENGINEERING CONNECTION

Surprise! All the package shapes hold about the same number of beans. Why do engineers decide to use one shape instead of another?

Engineers put a lot of thought into developing packaging for the products we buy—which materials to use, how much the package costs to make, is it attractive, does it protect the product, what is the impact on the environment, and much more. With more and more products being created, package engineering is a field with lots of opportunities!

DOMINO DIVING BOARD

Engineering Fields

- civil engineering
- mechanical engineering

Engineering Concepts & Skills

- role of failure
- modeling

Supplies

- set of dominoes (28 or more)
- thick, hard cover book
- ruler
- Domino Diving Board activity sign (Appendix A)

Advance Preparation Supplies

- tape

Advance Preparation

- Place the book and dominoes on a sturdy table that does not wobble or shake.

- Tape the ruler to the table with "zero" placed next to the book as shown below.

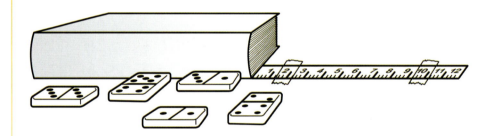

ENGINEERING CONNECTION

A **cantilever** is a structure that is connected to a support at one end and extends out beyond support on the other end. Engineers must design cantilevers to be structurally safe. The fixed end must have enough support, or weight, to hold up the weight of the extended end. Some examples of cantilevers are diving boards, balconies, and airplane wings.

Activity Steps

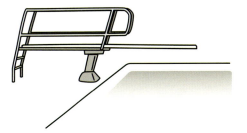

1. Build a ledge that "hangs out" over the edge of the book, like a diving board over a pool. **No dominoes can touch the table!**

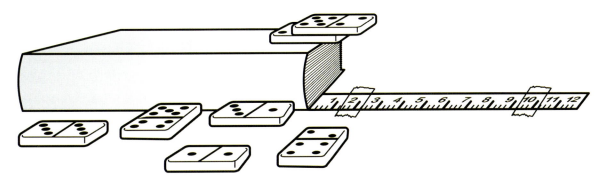

2. Watch the ruler to see how far your ledge "hangs out" before it collapses.

3. Improve your design and try again!

COLOSSAL CANTILEVER

The Grand Canyon Skywalk is a glass-bottom, horseshoe-shaped cantilever platform projecting 70 feet out over the cliff edge. It allows visitors to look straight down, nearly 4,000 feet below, to the bottom of the Grand Canyon.

GLUE IS THE CLUE

Engineering Fields

- *materials engineering*
- *chemical engineering*

Engineering Concepts & Skills

- *role of failure*
- *controlled experimentation and testing*

Supplies

- *4 colored index cards (4" x 6"), 2 of one color and 2 of another color*
- *2 plastic cups (12 oz. or larger)*
- *30 flat metal washers (1–1½" diameter)*
- *container to hold washers*
- *Glue is the Clue activity sign (Appendix A)*

Advance Preparation Supplies

- *glue stick*
- *ruler*
- *marker*
- *tape*

Advance Preparation

- Glue two index cards, one of each color, together like a sandwich. Allow them to dry thoroughly. Label both sides with the word "glued" across the top (see below).

- Label both sides of the remaining cards, one of each color, with the word "unglued" across the top.

- Draw a 2-inch diameter circle in the center on both sides of each pair of cards. This will indicate where the washers should be stacked to ensure a fair test.

- Set the two cups upside down on the table surface so that there are 4 inches between the tops of the cups. Tape them to the table to prevent them from moving.

Activity Steps

1. Stack the two **unglued** cards together and place them across the top of the two cups like a bridge. How many washers do you think the cards will hold before collapsing?

2. Carefully stack the washers in the circle on the top card, counting them as they are added. Keep going until the cards collapse.

3. Now, repeat with the **glued** cards. How many washers do you think the **glued** cards will hold?

4. Which pair of cards held more washers? Why?

ENGINEERING CONNECTION

The two **glued** cards stay together, becoming a single, thicker card with greater strength. The two **unglued** cards do not work as well together to hold up weight because they can bend and slide apart.

Gluing layers of material together is called **lamination**. Plywood is an engineered wood product that uses lamination. To make plywood, chemical engineers developed a special glue to hold many thin layers of wood together. This makes plywood stronger than solid wood of the same thickness.

Skateboards are made out of plywood so that they can be both strong and lightweight.

HAPPY FEET

Engineering Fields

- *biomedical engineering*
- *materials engineering*
- *industrial engineering*

Engineering Concepts & Skills

- *properties of materials*
- *reverse engineering*
- *engineering design process*
- *optimization/tradeoffs*

Supplies

- *5-6 single shoes used for activities depicted on Activity cards (soccer, ballet, football, basketball, running, hiking, construction, lounging at home, etc.)*
- *Activity Cards (Appendix B) —select one for each shoe type being used*
- *Design Requirement Cards (Appendix B)*
- *paper, approximately 12" x 24"*
- *Happy Feet activity sign (Appendix A)*

Advance Preparation Supplies

- *marker*
- *masking tape*

Advance Preparation

- Collect 5-6 single shoes representing the activities pictured on the Activity cards. Using too many shoes can be overwhelming, so it is recommended that you use no more than 6 shoes for this activity. Choose a variety of shoes that have differing design requirements, including some novel shoes, such as swim flippers, ballet slippers, or ice skates. Use clean, minimally worn shoes so that families do not mind handling them.

- Make a color copy of the *Activity Cards* onto cardstock and cut into individual cards. Select cards that match the 5-6 shoes you have collected.

- Copy the *Design Requirement Cards* page onto colored cardstock and cut into individual cards.

- If the cards will be used multiple times, laminating is recommended.

- Create a tabletop activity mat with paper. Divide the paper into three sections as shown in the illustration. Label one section "Activity." Label a second section "Shoe," and make it large enough to place a shoe inside. Label a third section "Design Requirements." Leave room for up to three Design Requirement cards. Tape the activity mat to the table.

Twelve different Happy Feet Activity cards are available in Appendix B.

ENGINEERING CONNECTION

Designing shoes for specific activities and conditions requires several kinds of engineers. Biomedical engineers help create a design that is good for the feet and can improve performance; materials engineers design and combine materials to meet special requirements (waterproof, soft, flexible, etc.); and industrial engineers design factory equipment to make the shoes.

Activity Steps

1. Match an Activity card with a shoe and place them both on the mat.

2. Select three Design Requirements that an engineer should consider when designing a shoe for this activity. Place these cards on the mat.

3. Repeat with a different Activity card and shoe.

4. Look at the shoes on your feet! What activity were they designed for?

INSPIRED BY NATURE

Engineering Fields
- *general engineering*

Engineering Concepts & Skills
- *engineering design process*
- *invention/innovation*

Supplies
- *set of Human Invention Cards (Appendix B)*
- *set of Nature's Inspiration Cards (Appendix B)*
- *sheet of paper (12" x 42")*
- *Inspired by Nature activity sign (Appendix A)*

Advance Preparation Supplies
- *marker*
- *masking tape*

How have sharks and termites helped engineers?

Advance Preparation

- Make color copies of the *Human Invention Cards* on cardstock and cut them apart. Laminating is recommended.

- Make two-sided color copies of the *Nature's Inspiration Cards* on cardstock and cut them apart. These cards are double-sided and must be copied so that the matching Human Invention information appears on the back of each card. Laminating is recommended.

- Divide the sheet of paper into two rows labeled "Human Invention" and "Nature's Inspiration" and tape the paper to the table. See the illustration below.

- Place all the cards on the table with pictures facing up.

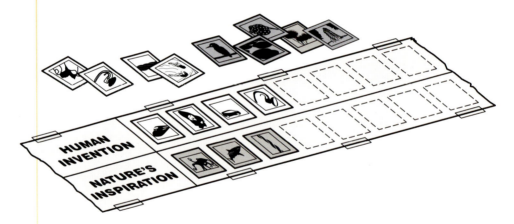

HUMAN INVENTION

Water Repellent Fabric

NATURE'S INSPIRATION

Lotus Leaf

Eight different Human Invention cards and eight matching Nature's Inspiration cards are available in Appendix B.

Activity Steps

1. Select a Human Invention card.

2. Next, select a Nature's Inspiration card to match this invention.

3. After all the cards have been matched and placed on the mat, read the backs of the Nature's Inspiration cards to find out how engineers are inspired by nature.

4. Can you think of other engineered inventions that might have been inspired by nature?

5. Remove the cards from the mat when you are finished.

 ENGINEERING CONNECTION

Sometimes engineers come up with a new idea for an invention by observing something in nature.

All of the inventions on the Human Invention cards have been inspired by something in nature. The story of the inventions can be found on the back of the Nature's Inspiration cards.

Next time you're outdoors, take some time to look around and wonder.

LEARNING FROM FAILURE

Engineering Fields

- general engineering

Engineering Concepts & Skills

- role of failure
- engineering design process
- open-ended problem-solving

Supplies

- plastic container at least 6-8" deep and 12" wide
- aluminum foil cut into 4" squares (approximately 5 squares per family)
- bowl of 50 pennies
- cookie sheet, plastic tray or hand towel
- paper towels or an extra hand towel
- water
- Learning From Failure activity sign (Appendix A)

Advance Preparation

- Cut the aluminum foil into 4" squares.

- Add water to the plastic container until it is half full.

- Place the container on a cookie sheet, plastic tray, or hand towel to catch splashes.

- Provide paper towels or an extra hand towel on the table for drying hands.

- *Event Tip*: Monitor this activity for water spills and slippery floors. If planning to use these materials for a future event, be sure to dry off all containers and pennies before storing.

ENGINEERING CONNECTION

Failure plays an important role in the design process. Engineers use failure to help them find better solutions. Often, engineers will test a design until it fails in order to see where improvements are needed. By testing your boat until it sinks and watching closely, you see where your design can be improved to keep the boat floating longer or to help it carry a heavier load of pennies.

Activity Steps

1. Create a boat out of **one** piece of aluminum foil and place it in the water. Predict how many pennies you think your boat will hold before it fails and sinks.

2. Place pennies in your boat gently, one-by-one. Watch the boat carefully as it gets close to sinking.

3. Can you change your boat design to hold more pennies? Try again using the same foil or **one** new piece.

4. What did you learn from watching your boat sink?

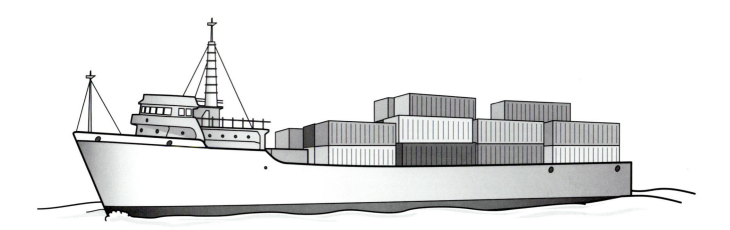

LET'S COMMUNICATE

Engineering Fields

- *general engineering*

Engineering Concepts & Skills

- *communication*
- *teamwork*
- *spatial ability*

Supplies

- *2 identical sets of 7 interlocking blocks of various sizes and colors (LEGO® blocks work well)*
- *2 resealable plastic bags or small containers to hold blocks*
- *2 small cardboard boxes (at least as big as an adult shoe box— approximately 8" x 5" x 12")*
- *Let's Communicate activity sign (Appendix A)*

Advance Preparation Supplies

- *marker*

As an engineer, how well can you explain your design to a builder?

Advance Preparation

- Divide the interlocking blocks into two identical sets so that each set includes the exact same blocks. Place each set in its own resealable plastic bag or small container.

- Prepare 2 cardboard boxes so that they are open on one side. Label the top of one box "Engineer" and the other box "Builder."

- Place the boxes on a table with the open sides facing opposite directions. Place one set of blocks in front of each box.

- Optional: if available, chairs can be placed on each side of the table.

- *Event Tip*: For young children, this activity can be done with fewer (and larger) blocks. Increasing the number of blocks can increase the challenge, but this will also lengthen the time it takes to complete the activity, making it less suitable for an event station.

Activity Steps

1. Face a partner so that you each have a box on its side and an identical set of blocks in front of you.

2. One of you will be the **engineer**. Design and build a structure inside your box. Do not allow your partner (the builder) to see your structure.

3. Next, have the **builder** try to build a copy of the engineer's structure in their own box by following the engineer's directions only. The builder cannot ask questions and you cannot look inside each other's boxes.

4. When finished, compare the two structures. How well did the engineer communicate to the builder?

ENGINEERING CONNECTION

Communication is harder than we think! To be successful, an engineer must be able to communicate effectively with others and work well on a team. Accurate and clear communication between all team members is key to a team's success.

MAKE IT LOUD!

Engineering Fields

- mechanical engineering
- biomedical engineering
- materials engineering

Engineering Concepts & Skills

- properties of materials
- controlled experimentation and testing

Supplies

- 5 cardboard paper towel tubes
- 5 strong rubber bands
- 5" x 5" square of each material: waxed paper, aluminum foil, paper towel, felt, plastic wrap
- Make It Loud! activity sign (Appendix A)

Advance Preparation

- Cover one end of each paper towel tube with a different square of material. Secure the material tightly with a rubber band.

- *Event Tip*: Have extra materials and rubber bands available in case a tube needs to be replaced.

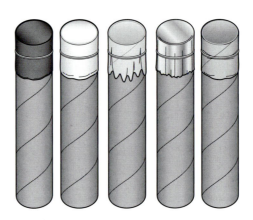

✴ ENGINEERING CONNECTION

Biomedical engineers design devices and procedures that solve medical and health related problems. These devices include artificial organs, artificial limbs, advanced imaging machines, and many other medical instruments and tools that help doctors diagnose and treat their patients.

One tool that is designed to make sounds louder is a **stethoscope**. Stethoscopes are used by doctors to make the sounds inside of people and other animals easier to hear. Listening to these sounds can help a doctor or veterinarian diagnose a possible medical problem. Stethoscopes are also used by mechanics to listen to the sounds made by machines or car engines.

Activity Steps

1. Hold the open end of one tube to your ear.

2. Have **someone else** tap lightly on the covered end. What do you notice?

3. Repeat using the other tubes, keeping the same level of tapping.

4. Which material makes the tapping sound louder? Which material makes the tapping sound quieter?

PICTURE THIS

Engineering Fields

- *general engineering*

Engineering Concepts & Skills

- *spatial visualization*

Supplies

- *2-3 pairs of scissors*
- *2-3 clear tape dispensers with tape*
- *copies of 4 different Design Pages—one of each page per family (Appendix B)*
- *4 pre-assembled 3D shape examples*
- *Picture This activity sign (Appendix A)*

Advance Preparation Supplies

- *marker*
- *4 pages heavy, white cardstock*

Advance Preparation

- Make enough copies of the four *Design Pages* to have one of each page per family. If possible, copy each design page onto a different color so that families can identify the designs by color, as well as by number.

- Make one copy of each design page on heavy, white cardstock. Cut out templates and use tape to assemble 3D shapes to use as examples.

- Use the marker to label each shape example with the appropriate name.

  - *Design Page #1* – Tetrahedron
  - *Design Page #2* – Cube
  - *Design Page #3* – Triangular Prism
  - *Design Page #4* – Rectangular Prism

- *Event Tip*: Provide additional pairs of scissors and tape dispensers according to the number of families expected to attend.

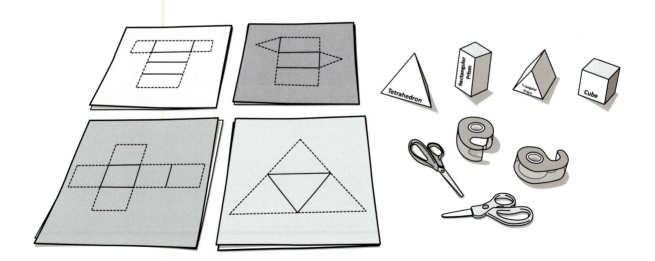

Activity Steps

1. From the four examples, choose a 3D shape you would like to make.

2. Select a "Design Page" you think will make this shape.

3. Cut out your shape by cutting along the dotted lines only.

4. Fold along the solid lines, then use tape to complete your 3D shape.

5. Does your 3D shape look the way you thought it would?

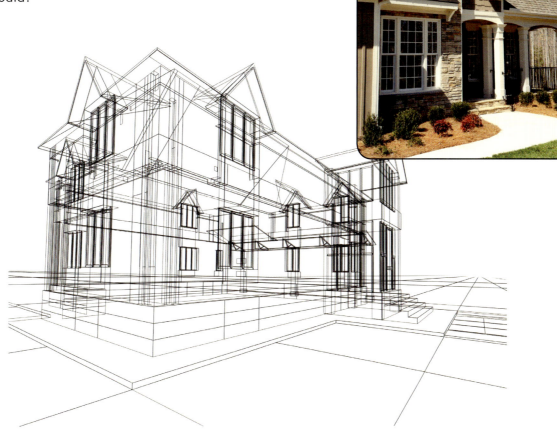

ENGINEERING CONNECTION

When engineers design something, they usually draw a two-dimensional (2D) version on paper or with a computer before assembling a three-dimensional (3D) model. The ability to imagine or "see" how a 2D picture can become a 3D object is called **spatial visualization**, a very useful skill in engineering!

SHIFTING SHAPES

Engineering Fields

*   *civil engineering*

Engineering Concepts & Skills

*   *modeling*
*   *open-ended problem-solving*

Supplies

*   *poster board or lightweight cardboard (empty cereal boxes work well)*
*   *30 brass paper-fasteners in a bowl or container*
*   *Shifting Shapes activity sign (Appendix A)*

Advance Preparation Supplies

*   *hole punch*
*   *scissors*

Advance Preparation

*   Cut the cardboard into 1"x 5" strips and punch a hole near each end. Make about 30 strips.

*   Make two triangles and two squares by fastening strips together with brass paper-fasteners.

*   Place materials, additional strips, fasteners, and sample shapes on the table as shown in the illustration below.

*   *Event Tip*: You may need to disassemble family-made shapes periodically so that only pre-made triangles and squares are available for families when they approach this activity.

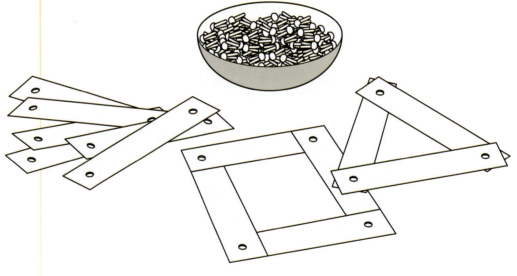

ENGINEERING CONNECTION

When engineers design a structure, it needs to be stable and keep its shape. Engineers often add strength and stability to the structures they design by using materials in the shape of a triangle. Triangles don't twist, bend, or collapse easily. If you look closely, chances are you will see lots of triangle shapes in the structures around your community.

Activity Steps

1. Select a square and a triangle. Gently try to move the sides of each shape up and down.

   - Which one is more rigid and keeps its shape?
   - Which one shifts and changes shape easily? Can you add another strip to this shape to make it more stable?

2. Try making a 5-sided shape. Does it shift and change its shape easily? What can you do to make it more stable?

3. Before you leave, please take apart any extra shapes, or added fasteners and strips, leaving only the original squares and triangles at the table.

SHOWERHEAD SHOWDOWN

Engineering Fields

- environmental engineering
- mechanical engineering

Engineering Concepts & Skills

- optimization/tradeoffs
- sustainability

Supplies

- 2 clear 9 oz. plastic cups that are short (3") with a wide open rim
- plastic container at least 8" wide, 6-8" deep
- water
- hand towel
- paper towels
- small trash container for used paper towels
- Showerhead Showdown activity sign (Appendix A)

Advance Preparation Supplies

- large push pin with plastic top
- sharpened pencil
- permanent marker

Suggested Size for Holes

Small •

Large ●

Advance Preparation

- **Make Showerhead #1:** Using a plastic-top push pin, carefully punch 15 small holes (see suggested size below) in the bottom of one plastic cup. Punch the holes **from the inside out** while gently twisting the pin tip back and forth. Be sure to keep the holes small, but also large enough for water to pass through. Using the permanent marker, label this cup "Showerhead #1."

- **Make Showerhead #2:** Carefully punch 5 larger holes (see suggested size below) in the bottom of a second cup by first using the push pin and then making the holes wider with the tip of a pencil. Only push the pencil in about halfway up the point, not all the way to the full diameter of the pencil. Punch the holes **from the inside out**, twisting the pencil back and forth. Using the permanent marker, label this cup "Showerhead #2."

- Fill the plastic tub with water to a depth 1-2" **deeper** than the height of the cups.

- Place a hand towel under the plastic tub to soak up splashes and some paper towels on the table for drying hands.

- Test the cups (showerheads) to make sure that the water is flowing adequately through the holes and that Showerhead #1 takes noticeably longer to empty than Showerhead #2. Adjust size of holes if necessary.

- *Event Tips*: Place the small trash container beneath the table for used paper towels. Check on this activity area periodically to wipe up any water that may have splashed onto the floor.

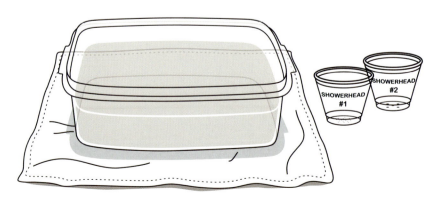

Activity Steps

1. Look at the bottom of the two cups. How are they different? Which one do you think will drain water the fastest?

2. Do the 5-second shower test. Start by holding the cups upright by their open rims.

3. Lower both cups straight down all the way under the water so they fill to the very top (see below).

4. Lift both cups out of the water **at the same time**, holding them over the tub to drain. Count to 5 slowly.

5. Which cup (showerhead) uses the least amount of water for a 5-second shower? Which showerhead design would help you save water at home? Why?

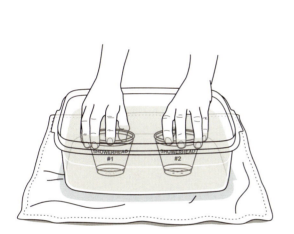 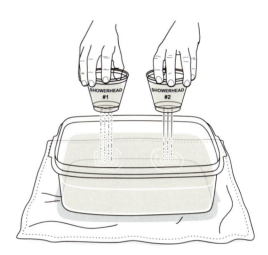

ENGINEERING CONNECTION

In the United States, showering consumes about 20% of the total water used by a family in an average home. Older showerheads have fewer but larger holes for water flow. Newer "low flow" showerheads are designed with a greater number of smaller holes. "Low flow" showerheads are engineered to use less water, save energy, and still provide a nice shower.

SOLID GROUND

Engineering Fields

- geological engineering
- civil engineering

Engineering Concepts & Skills

- optimization/tradeoffs
- controlled experimentation and testing

Supplies

- 3 plastic containers (quart size)
- 1½ cups of each of the following loose earth materials: natural crushed rock with rough, angular edges (¼ inch pieces), natural rounded pea-size gravel with smooth, tumbled edges (e.g., aquarium gravel), and clean natural sand
- 3 large plastic LEGO®/DUPLO® blocks or other same size blocks
- Solid Ground activity sign (Appendix A)

Advance Preparation Supplies

- masking tape
- marker

Advance Preparation

- Fill each of the three plastic containers with 1½ cups of a different earth material.

- Use masking tape and a pen to label the containers—"Rough Gravel," "Smooth Gravel," and "Sand."

- Place a large LEGO®/DUPLO® block on top of the earth material in each container.

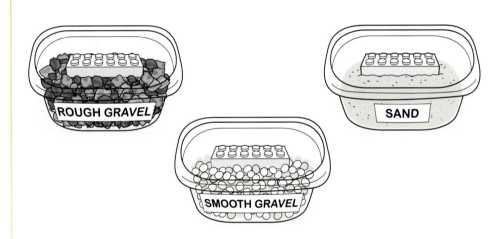

⭐ ENGINEERING CONNECTION

Geological and civil engineers test different types of earth materials to determine a good place to build. Materials such as crushed rock, where the pieces are rough and do not slide against each other easily, can lock together and support more weight or pressure. This is a good material to use as a foundation under a building or a road.

Smooth, rounded particles, like sand or rounded pea gravel, can easily slide against each other and move easily around other objects. This creates a surface that is not a good foundation for heavy construction, but can be good for absorbing impact and cushioning a fall.

A house should be built on a firm foundation.

A playground also needs a firm foundation, but the material covering the surface needs to cushion a fall.

Activity Steps

1. Place the block **on top** of the earth material.

2. Place two fingers in the middle of the block and press down. What happens?

3. Repeat with the other two earth materials. Does the block push easily into each material?

4. How are the materials different? What do they feel like?

5. If you were an engineer, which material would you use to provide a firm foundation for a house? Which material would you use under a playground, to cushion a fall?

SOUNDPROOF PACKAGE

Engineering Fields

- *acoustical engineering*
- *materials engineering*
- *package engineering*

Engineering Concepts & Skills

- *open-ended problem-solving*
- *engineering under constraints*
- *properties of materials*

Supplies

- *2 identical small containers with tight-fitting lids (film canister size)*
- *10 pennies or dry beans*
- *2 identical one-quart plastic containers with lids*
- *A variety of testing materials, enough of each to fill two one-quart containers: felt, foam cushion, cotton balls, bubble wrap, newspaper, tissue paper, paper towels, quilt batting, netting, foil, Styrofoam packing peanuts, etc.*
- *Soundproof Package activity sign (Appendix A)*

Advance Preparation Supplies

- *marker*
- *tape*

> **Can you engineer a package that reduces noise?**

Advance Preparation

- Gather testing materials, including items that will muffle sound as well as items that will not muffle sound. Be sure to have enough of each material to fill the inside of a one-quart container.

- Prepare two noisemakers by placing 5 pennies or dry beans inside each small container, taping over the sealed lids, and using the marker to label each one "noisemaker."

- *Event tip:* Place loose materials into separate bins to keep the activity station organized.

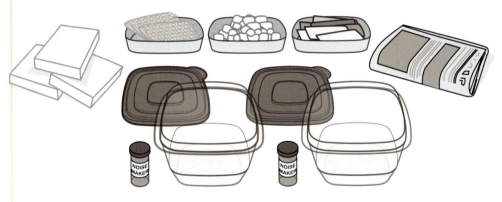

Activity Steps

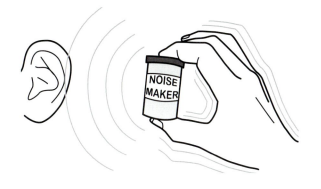

1. Shake a noisemaker to hear the noise it makes.

2. Design a package for the noisemaker that will muffle the noise. Use one or more of the materials provided and package the noisemaker in the plastic container. Attach the lid.

3. Test your package by shaking it. Can you still hear the noisemaker?

4. Try again with different materials. Which materials make the best soundproof package?

5. Please take apart your package and separate the materials when you have completed your tests.

ENGINEERING CONNECTION

Noise is unwanted sound. Engineers design products and materials to protect the human ear from loud noises and reduce or remove unwanted sounds.

People wear ear protection when working around loud noises.

THRILL SEEKERS

Engineering Fields

- *mechanical engineering*

Engineering Concepts & Skills

- *teamwork*
- *modeling*

Supplies

- *8 foot length of flexible clear plastic tubing at least ¾ inch diameter, available at hardware stores; best to buy bulk from large spool— precut rolls tend to kink*
- *5 very small marbles (diameter must be less than ½ inch to fit in tube)— called "mini-marbles" at educational supply stores*
- *plastic wrap*
- *rubber band*
- *small plastic cup or container to hold marbles*
- *Thrill Seekers activity sign (Appendix A)*

Can you engineer a good roller coaster design?

Advance Preparation

- Use a rubber band to secure a piece of plastic wrap over one end of the tubing. This will help to catch the marble.

- *Safety Note:* Small marbles may pose a choking hazard to children under age 3.

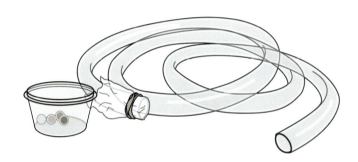

ENGINEERING CONNECTION

Mechanical engineers designing a roller coaster must balance the thrill of the ride with safety. They use the science of how things move to design hills, loops, twists, and turns that give riders a safe but thrilling ride!

THE ORIGINAL 'COASTER'

The idea of 'coasting' for fun comes from ice slides popular in Russia in the 17th century. Huge wooden structures with a thick sheet of ice allowed people to climb a stairway, get on a sled, and then careen down a steep ramp and up an opposite ramp, going back and forth before eventually stopping in the middle.

Activity Steps

1. Work as a team to hold the tubing in a roller coaster shape.

2. Put a marble into the top of the tube and watch it ride. Did it make it to the end?

3. How many loops and turns can you add to make the most thrilling ride and still get the marble to the end?

TUMBLING TOWER

Engineering Fields

*   *civil engineering*

Engineering Concepts & Skills

*   *role of failure*
*   *modeling*

Supplies

*   *14 empty toilet paper tubes* **of the same length**
*   *3 squares of 12" x 12" corrugated cardboard*
*   *Tumbling Tower activity sign (Appendix A)*

Activity Steps

1. Use 14 tubes and 3 cardboard squares to build this tower.

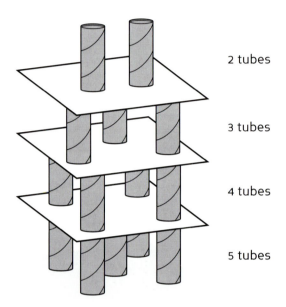

2 tubes

3 tubes

4 tubes

5 tubes

2. Following the rules below, take turns removing one tube at a time from the tower without letting the 3 cardboard platforms fall.

    *   You may use both hands.
    *   You may touch the cardboard platforms only when removing or moving a tube.
    *   You may change the position of the remaining tubes.

3. What steps did you take to keep the tower from tumbling? Why did the tower eventually fall?

ENGINEERING CONNECTION

Engineers often test a structure's strength until it fails and then try to figure out why the failure happened. The weight that a structure supports is called its **load**. In order for a structure to be stable, the load must be balanced. If a structure is changed in some way, the load may need to be rebalanced to maintain stability.

WHAT DO ENGINEERS DO?

What do *you* think engineers do?

Engineering Disciplines

- *general engineering*

Engineering Concepts & Skills

- *career awareness*

Advance Preparation

- Copy the *Activity Statements* and cut them into individual cards.

- Line the cups up on a table and tape an activity statement card in front of each cup. Place the bowl of beans in front of cards.

- *Event Tip:* After everyone has completed this activity, gather them together as a large group. Discuss some of the common misconceptions about what engineers do and the wide range of engineering fields that exist. See "The World of Engineering" chapter for more information on engineering fields.

Supplies

- *10 plastic cups (12 oz. or larger)*
- *Activity Statements (Appendix B)*
- *2-3 cups of large dried beans in a bowl (kidney, lima, navy, etc.)*
- *What Do Engineers Do? activity sign (Appendix A)*

Advance Preparation Supplies

- *tape*
- *scissors*

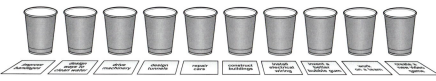

Activity Steps

1. Read the statement on each card.
2. Place one bean into each cup that you think describes what an engineer does.

☀ ENGINEERING CONNECTION

Engineers are creative problem-solvers who play a vital role in society. People often believe that engineers fix or build things and that they only work on buildings, bridges, and cars. Actually, engineers are the people who **design** these and thousands of other products. There are many different engineering fields.

- Some engineers develop new technologies to help sick or injured people feel better, such as new medical machines, medicines, and devices to help people walk, see, or hear.

- Some engineers develop solutions for everyday challenges and design products we use at home, such as can openers, refrigerators, computers, and packaged food.

- Some engineers even help design solutions to the world's biggest problems—like the need for clean drinking water or developing new energy sources.

WHO ENGINEERED IT?

Engineering Fields

· *general engineering*

Engineering Concepts & Skills

· *teamwork*

Supplies

· *set of Engineered Product Cards (Appendix B)*
· *set of Engineer Cards (Appendix B)*
· *sheet of paper, 20" long x 12" wide*
· *Who Engineered It? activity sign (Appendix A)*

Advance Preparation Supplies

· *marker*
· *tape*

> ### How many engineers does it take to design a light bulb?

Advance Preparation

· Make color copies of the 2-sided *Engineered Product Cards* on cardstock. Cut each page into four individual cards. Laminating is recommended if the cards will be used on multiple occasions.

· Make color copies of the *Engineer Cards* on cardstock. Cut each page into four individual cards. Laminating is recommended if the cards will be used on multiple occasions.

· Use the 20" x 12" paper to prepare a game board, providing a 4" left side column for the Engineered Product cards and an 8" right side column for the Engineer cards. Label the columns as shown.

· Place the Engineered Product cards in a stack with the product photo facing up. Place the Engineer cards in a stack with the photos facing up.

☀ ENGINEERING CONNECTION

Products and structures don't just "happen"! They are imagined, designed, and engineered by people trained in an engineering field.

Most products are actually designed by a team of engineers from different engineering fields. For example, to design and produce an automobile requires at least five different kinds of engineers working together: mechanical, electrical, computer, materials, and biomedical engineers.

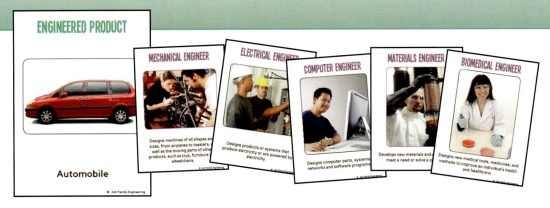

Activity Steps

1. Deal the Engineer cards out to family members until all cards are distributed.

2. Choose an Engineered Product card and place it on the game board.

3. Have each family member choose the Engineers from their hand that they think helped design this product and place these cards on the game board.

4. Turn over the Engineered Product card to check your answers.

5. Deal the Engineer cards again and repeat with another Engineered Product card.

6. Please leave the two sets of cards in separate stacks on the table when you are finished.

WRAP IT UP!

Engineering Fields

- materials engineering

Engineering Concepts & Skills

- properties of materials
- controlled experimentation and testing

Supplies

- cotton swabs in a bowl or container (approximately 5 per family)
- 3" x 3" squares of five different materials (one set per family): wax paper, paper towel, aluminum foil, newspaper, and heavy plastic garbage bag material
- small, shallow clear plastic container (not larger than a 6" x 6" base)
- water
- cookie sheet or plastic tray
- small table top container for waste materials
- Wrap It Up! activity sign (Appendix A)

Advance Preparation Supplies

- scissors
- dark-colored washable marker

Advance Preparation

- Cut and sort 3" squares of wax paper, paper towel, aluminum foil, newspaper, and heavy plastic garbage bag material so that there is enough for at least one square of each material per family.

- Place water in the clear plastic container so that it is only 1/2 inch deep (just deep enough to cover the tip of the cotton swab).

- Color the water by swirling open tip of washable marker in the water. Using washable marker instead of food coloring will allow for splashes on clothing to be washed out.

- Place activity materials on the table as shown in illustration below.

- Label the "waste" container.

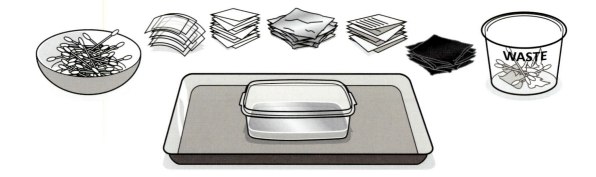

Activity Steps

1. Which materials will keep a cotton swab dry?

   - Select one square of any material, wrap it around the end of a cotton swab, and hold it in place with your fingers.
   - Dip the covered tip of the cotton swab into the water and count to 10, slowly.
   - Lift the cotton swab out of the water and remove the square of material. Is the tip of the cotton swab dry or wet?

2. Test the other materials with a dry cotton swab. Which material would you use to protect your house from water damage?

ENGINEERING CONNECTION

When engineers design a building, it is important for them to know how different materials react to water. This knowledge will help them choose water resistant construction materials or add a barrier layer that will keep the water out altogether. In house designs, engineers usually use black tar paper or plastic sheeting to act as a water barrier to protect the wood underneath.

YOUR FOOT, MY FOOT

Can engineers use their feet for measuring?

Engineering Fields

- *general engineering*

Engineering Concepts & Skills

- *communication*

Advance Preparation

- Use masking tape to make a 6-foot long line of tape on the floor.

Activity Steps

1. Have each family member measure the length of the tape line using their own feet. Do this by walking heel to toe along the tape and counting the number of "foot" lengths from end-to-end.

2. Did everyone get the same number?

3. Why is it important to use standard units for measuring?

Supplies

- *measuring tape*
- *masking tape*
- *Your Foot, My Foot activity sign (Appendix A)*

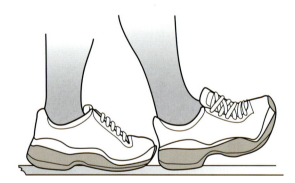

✴ ENGINEERING CONNECTION

When people talk to each other, speaking the same language helps them to understand each other. This is also true with measurement. It is important that a unit of measurement (such as one foot) means the same thing to everyone. That is why we measure length with a ruler that has a standard length for one foot, instead of letting people measure with their own feet that are all different sizes.

Accurate measurement is essential in engineering. Engineers use accurate measurements to help them draw a design, construct models, or give instructions to the people who will create or build a product the engineer has designed.

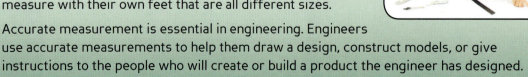

ENGINEERING CHALLENGES

7

Engineering Challenges are longer, more in-depth activities that engage families in the process of engineering and introduce them to a range of engineering concepts and skills, as well as to a variety of engineering fields. These activities are generally designed to be completed in 30-45 minutes with family members working together as a team. At a Family Engineering event, volunteer facilitators will lead an Engineering Challenge activity for a group of families.

Many of these activities are "design challenges" in which families work together to design something that meets specific requirements, such as looking or functioning in a specific way. There may be certain constraints, such as limited time or limited materials. During a design challenge, families are introduced to and encouraged to use the engineering design process as they tackle the challenge. The engineering design process is described in detail in "The World of Engineering" chapter. Engineering Challenge activities provide fun and engaging ways of introducing general engineering concepts such as problem-solving, teamwork, and innovation. Engineering Challenge activities also introduce families to the many different fields of engineering, and to how engineering affects our daily lives.

Activity	Page Number	Estimated Activity Time	Advance Preparation Time	Difficulty Level	Activity Type	Engineering Fields
Artistic Robots	86	40-50	■	●	DC	Mechanical, Electrical
Assembly Line	90	30-40	□	◐	DC	Industrial, Systems
Blast Off!	94	40-50	■	◐	DC	Aerospace
Brain Saver	98	35-45	◩	○	DC	Biomedical, Mechanical, Materials
Bright Ideas	102	40-50	■	◐	DC	Electrical
Create a Critter	106	45-55	■	●	DC	Mechanical
Engineering All Around	112	25-35	◩	◐	G	General
Engineering Charades	114	20-30	□	○	G	General
Five Points Traffic Jam	116	30-40	◩	◐	DC	Civil, Systems
Give Me a Hand	119	30-40	◩	◐	DC	Biomedical, Mechanical
Hot Chocolate Machine	122	40-50	■	●	DC	Chemical, Mechanical
Launcher	126	30-40	◩	◐	DC	Aerospace, Mechanical
Mining for Chocolate	130	35-45	□	◐	DC	Mining, Environmental, Geological
Stop and Think	134	20-30	□	○	G	General, Materials
Team Up!	137	20-30	□	○	G	General

Advanced Preparation Time
□ Very little
◩ Moderate amount
■ More extensive

Difficulty Level
○ Simplest
◐ Moderate difficulty
● More complex challenges

Activity Type
G - General Engineering
DC - Design Challenge

Choosing Engineering Challenges

When choosing Engineering Challenge activities for an event, think about the time available for facilitated activities, the format of the event (are activities being facilitated in one large group or conducted concurrently in different rooms with different facilitators), and the interests and needs of your participating families. Will most of the children present be about the same age? Will activities with a lot of reading be suitable for the participating families? Can you assemble the necessary supplies if you expect a large number of families to attend?

> *Choose activities that represent different engineering fields and/or highlight different aspects of engineering.*

One of the goals of Family Engineering is to introduce families to the many different career paths in engineering, thereby broadening their understanding of who engineers are and what they do. When planning a Family Engineering event, choose activities that represent different engineering fields and/or highlight different aspects of engineering. Use the "Engineering Challenges At-A-Glance" chart to help you select activities that meet your event needs and also give families a broad exposure to engineering. Remember to consider which Opener activities families will experience during the event and select longer activities that either complement these initial experiences or add to the diversity of engineering fields being introduced.

Listed below are some suggested activity combinations for two different event formats. Each combination offers a mix of activity types, levels of difficulty, and introduces families to a variety of engineering fields.

Single Large Group Format—When facilitating multiple Engineering Challenge activities in one room with a large group of families, you will need to consider the time it takes to transition from one activity to another. It is helpful to have extra volunteers available to assist with dispensing supplies and cleaning up. See the "Organizing a Family Engineering Event" chapter for additional tips on event planning and facilitation.

Available Time for Engineering Challenge Activities	Suggested Activity Combinations		
60-75 minutes	Team Up! Assembly Line	or	Engineering Charades Brain Saver
75-90 minutes	Stop and Think Artistic Robots	or	Engineering All Around Hot Chocolate Machine
90-100 minutes	Give Me a Hand Create a Critter	or	Mining for Chocolate Bright Ideas

Multiple Smaller Groups Format—If the Family Engineering event is formatted with smaller groups of families rotating from one room to the next, make sure that the activities have the same estimated time for completion. This will help to keep everyone on the same rotation schedule. This format requires at least one volunteer activity facilitator for each activity room, and at least one activity assistant is recommended. See the "Organizing a Family Engineering Event" chapter for additional tips on event planning and facilitation.

Length of Concurrent Activity Sessions	Suggested Activities for Concurrent Sessions
30-40 minutes	Assembly Line Give Me a Hand Five Points Traffic Jam Launcher
40-50 minutes	Blast Off! Brain Saver Mining for Chocolate Stop and Think & Engineering Charades
50-60 minutes	Bright Ideas Engineering All Around & Team Up! Create a Critter Hot Chocolate Machine

ARTISTIC ROBOTS

Family members work as a team using their imagination and simple materials to design an "Artistic Robot" that will create colorful drawings all on its own!

Engineering Fields

- *mechanical engineering*
- *electrical engineering*

Engineering Concepts & Skills

- *engineering design process*
- *teamwork*
- *open-ended problem solving*
- *role of failure*
- *invention/innovation*

Estimated time: 40-50 minutes

General Supplies

- *multiple large sheets of newsprint or butcher paper*
- *masking tape (one roll for every 2-3 families)*

Supplies Per Family

(one set for **each** family or small group)

- *Artistic Robots Design Challenge (Appendix C)*
- *In a gallon-size resealable plastic bag*
  - ☐ *1.5 volt motor with wire leads (found at hobby shops or on the Internet)*
  - ☐ *AAA or AA battery*
  - ☐ *large plastic cup (thin plastic works best)*
  - ☐ *3 water-based markers (not permanent markers)*
  - ☐ *2 drinking straws*
  - ☐ *2 rubber bands*
  - ☐ *2 craft sticks*
  - ☐ *2 paper clips*
  - ☐ *eraser*
  - ☐ *pencil*
  - ☐ *2-3 sheets of 11" x 17" white paper*

Advance Preparation

- Make copies of the *Artistic Robots Design Challenge* (one per family).

- Pre-package family supplies into resealable plastic bags for easy distribution. The rolls of masking tape can be shared between family groups.

- Tape down several large sheets of newsprint or butcher paper on an area of the floor to provide a place for families to test their robots. If space allows, it is recommended that you make this area large enough for the entire group to gather around the perimeter to watch their different robots in action.

- *Event Tip:* Be aware that a washable marker may spatter ink if it is moving in a quick, flipping action as part of a robot.

Activity Steps

1. Ask family members to raise their hands if they have any toys at home that move, light up, or make sounds when you push a button or flip a switch. What powers these toys? (batteries) Explain that these toys are using energy from the batteries to power their activities. To make these kinds of toys, a mechanical engineer designs an object that will move, make sounds, or complete a task. Then the mechanical engineer needs the help of an electrical engineer who designs how to power the toy and makes it work. Often engineers get to work together and be creative!

2. Explain to families that they will be working together as a team of engineers and using their creativity to design and build an artistic robot. Show families a motor and a battery and let them know that these will be two important parts of their robot design—the parts that will make the robot move. Ask if anyone knows how to make the energy flow from the battery to the motor. Have a volunteer come up and show how to attach the wires to each end of the battery. Place a piece of masking tape on the spinning rod so that the group can see that the motor is moving when connected to the battery. Remind families that

tape should be used to connect the wires to the end of the battery and that they should not hold the bare wire on the battery with their fingers.

3. Distribute the *Artistic Robots Design Challenge* to each family and have them review the challenge and requirements.

4. Quickly review the engineering design process with the families and encourage them to use it while designing their robots. See "The World of Engineering" chapter for more information about the engineering design process.

5. Tell families that they will have 5-10 minutes to talk about the challenge and brainstorm ideas for a solution. Show the group the materials that they will receive. Point out that there will be many possible solutions to this challenge and say that they can use any items from the bag that they wish. Let families know that these robots will not be going home with them and that the supplies will be returned to the bags at the end of the activity.

6. Tell families that the "Ask" step is an important part of the design process. Asking questions can help to define the problem and discover any constraints or requirements, such as what materials can be used or what the robot needs to be able to do. Ask families if they have any questions about the challenge before they begin.

7. Distribute the supply bags and have families first remove only the pencil and a piece of paper. Ask them to begin brainstorming and sketching possible designs.

8. After 5-10 minutes, announce that families will now have 15-20 minutes to assemble and test their robot. Explain that families can use the paper to record evidence of their robots' artistic capabilities. Caution them to leave the marker caps on until their robot is on the paper and ready for a test. Encourage families to test their robot often and make adjustments to their design as needed.

Example Artistic Robots: These examples of designs are for the activity facilitator and should not be shared with families prior to engaging in the activity.

9. If some families seem discouraged by attempts that are not working, encourage them to look around at what others are doing to see if that gives them new ideas. Explain that engineers often look at what others have done to get inspiration and try to improve on their own ideas.

10. As families work on their robots, circulate among them and ask them to share problems they are having, remembering a few situations for use in the final step.

11. When families have finished constructing and testing their robots, gather everyone around the floor area covered with paper. Have all the robots working at the same time on the paper so that everyone can see the variations in robot design as well as the "art" that each robot draws. After a few minutes of robot activity, have families stop their robots.

12. Ask families to share a problem they encountered while building their robots and how they eventually solved the problem. Explain that, to engineers, failure is an important tool. They don't give up—they try a new idea. Share this story about Thomas Edison: when asked about all his failed attempts to create an incandescent light bulb, he was said to reply, "They're not failures. They taught me something that I didn't know. They taught me what direction to move in."

13. Ask families to dismantle their robots and return the re-usable supplies to the resealable plastic bag.

ENGINEERING CONNECTION

We are surrounded with examples of technological creativity everyday. Light bulbs, televisions, motor vehicles, and the buildings in which we live and work are all the result of the creative application of technology. Engineers often use a combination of technical knowledge and creative problem solving to design innovative products, including robots!

Robots are put to work in many different arenas, including the manufacturing industry, the military, space exploration, transportation, and the medical field. They often do jobs that are difficult, dangerous, or dull, such as lifting heavy objects, handling chemicals, or performing assembly work. Robots can also enter environments that are inhospitable, out of reach, or dangerous for humans, thereby allowing humans to remotely explore new frontiers. Robots can go inside of the human body during delicate surgery, explore at great depths in the ocean, work in radioactive zones, and take pictures on the surface of Mars.

A robotic rover explores the surface of Mars.

A PET ROBOT

AIBO®, a robotic dog, was designed and manufactured by Sony, starting in 1999. AIBO® is a social robot, one that can interact and communicate with humans. It is able to walk, speak, "see" its surroundings, and respond to commands in both English and Spanish. AIBO® can learn and develop from a 'puppy' to a 'mature' dog by interacting with its environment. This robotic pet can even express emotions!

Extensions

▶ Challenge families to use materials at home to create an artistic robot. Can they create a robot that will make specific designs, such as circles or dotted lines? What other things might their artistic robot do?

▶ Suggest that families research robots on the Internet and discover how robots are being used today.

ASSEMBLY LINE

Engineering Fields

- *industrial engineering*
- *systems engineering*

Engineering Concepts & Skills

- *reverse engineering*
- *optimization/tradeoffs*
- *communication*
- *teamwork*
- *systems*

Estimated time: 30-40 minutes

General Supplies

- *timer, stopwatch, or clock with second hand*

Supplies Per Family

(one set for **each** family or small group)

- *Assembly Line Data Sheet (Appendix D)*
- *In a gallon-size resealable plastic bag*
  - ☐ *5-6 identical retractable ballpoint pens that are easy to disassemble & re-assemble (Note: PILOT EasyTouch® Retractable Pens work well because they can be disassembled into 6 interchangeable parts and require 5 steps for assembly.)*
  - ☐ *6 small paper plates (teams should have at least as many paper plates as pen parts)*
  - ☐ *pencil*

How quickly can a team of workers assemble a product? Family members will reverse engineer a ballpoint pen to discover how it was designed and assembled. Then, working as a team, they will create an assembly line to optimize the process of correctly re-assembling all their ballpoint pens in the least amount of time.

Advance Preparation

- Make copies of the *Assembly Line Data Sheet* (one per family).
- Disassemble and reassemble one retractable ballpoint pen to determine the number of parts and the number of steps that it will take to put the pen back together.
- Pre-package team materials for easy distribution. In a resealable bag, place ballpoint pens, paper plates, pencil, and a copy of the *Assembly Line Data Sheet*. The number of plates should match the number of pen parts when one pen is disassembled.
- *Event Tip:* Have additional pens available to quickly replace pens that break or lose a part during the activity.
- *Safety Note:* Remind families with small children to be sure that small pen parts are not put in children's mouths.

Activity Steps

1. Organize families into teams of 5-6 individuals. Be sure to have a mix of adults and children in each group.
2. Distribute the following materials to each family: 1 ballpoint pen per team member, 6 small paper plates, *Assembly Line Data Sheet*, and pencil.
3. Have the families distribute one ballpoint pen to each family member, and place the paper plates in the middle of their table to store pen parts.

4. Have everyone carefully disassemble their pens to determine the total number of separate parts. There should be 5-6 individual parts—e.g. the ink cartridge, the spring, the top barrel of the pen, the bottom barrel of the pen, the push button, and the ratchet. The long ink cartridge can be used to push other parts out of the pen if necessary.

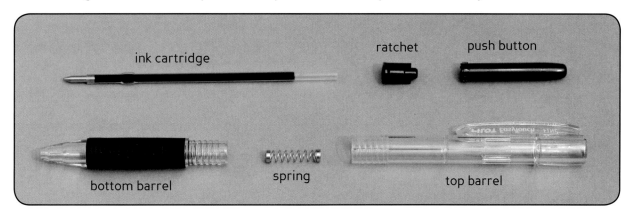

5. Ask families to discuss what each part is used for, and how it contributes to the way the pen works. Explain that this process is a form of **reverse engineering**—taking apart an object to see how it works. Next ask families to think about what steps they will need to take to re-assemble their ballpoint pens. Allow time for families to practice re-assembling their ballpoint pens.

6. Announce to the group that their first team challenge will be to see how fast each member of the group can re-assemble their own ballpoint pen. Ask everyone to disassemble their pens again and place all like parts in separate paper plates (parts bins) in the middle of their table. For example, all the ink cartridges will be together on one paper plate.

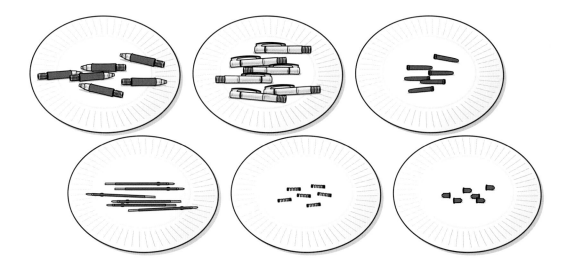

7. Have teams identify one family member to serve as their recorder. Inform the teams that you will call out the time every 5 seconds as teams get close to finishing. The recorder should write down the time closest to the moment when all the team's pens are re-assembled in working order.

8. When all pens have been disassembled, start the assembly process by saying "GO!" Begin timing the process using the stopwatch or clock. Each individual on a team should get parts from the parts bins in the middle of the table and re-assemble his/ her own pen. Each team member should only work on one pen. As families get close to completing the task, announce the time every five seconds until all of the pens are assembled. When all the members on a team have completed assembling their pens, the team should raise their hands. The recorder for that team should then record the time of completion next to "Individual Assembly" on their data sheet.

9. Allow time for families to discuss some of the challenges of having each member of the team re-assemble a pen on his or her own.

10. Announce to the group that their next challenge will be to see if they can improve their family's combined assembly time by working together in an organized assembly line. Allow about 10 minutes for each family to discuss and design their assembly line process. Remind them that the goal is still to re-assemble all of their pens, but this time each team member will engage in a specific assembly task.

DESIGNING A BETTER PROCESS

Model T assembly line.

American car manufacturer Henry Ford (1863-1947) developed the assembly line technique for large-scale production. His business was successful, in part, through the continuous improvement of his assembly line processes. Ford and his workers installed the first conveyor belt-based assembly line in his car factory around 1913-1914. The assembly line reduced the production costs for automobiles significantly by reducing the time it took to make a car. Using his engineered assembly line, Ford's famous Model T automobile was eventually assembled in ninety-three minutes. As a result, the automobile became more affordable to the average American.

11. Once all the members on a team know their roles, ask everyone to disassemble their pens again. They may organize the individual parts in whatever way they choose for their particular assembly line process. As before, start the assembly process by saying "GO!" and start timing the process. Keep announcing the time until all families have completed assembly. Have each recorder note their "Assembly Line Trial #1" completion time on the data sheet.

12. Take a moment to find out from the various families if this assembly line process was slower or faster than their earlier individual assembly time. Ask the families to discuss as a team what they observed and experienced with the first trial of their assembly line. Did every step in the process work the way they had imagined? What could they do to improve their process, resulting in even faster assembly? Explain to the group that working to improve their assembly line process is a form of **optimization**—making adjustments to a process to increase its efficiency or outcome.

13. Give the families 5 minutes to improve their process, if they wish, before disassembling their pens again for a second trial. Start the assembly process with another "Go!" and announce times as before. When all families have recorded their "Assembly Line Trial #2" time, ask if their times improved with the second trial. As time allows, have a few families share with the whole group the various approaches they used to optimize their assembly lines.

14. Explain to the group that engineers don't just design *things*—they also design systems and processes, such as creating the most efficient method for assembling a ballpoint pen.

Extensions

▶ Challenge families to try designing and optimizing an assembly line at home that accomplishes a routine and repeatable task such as making multiple sandwiches for lunch, folding and storing laundry, or setting the table for dinner.

ENGINEERING CONNECTION

Industrial engineers work to design the best possible way of doing or making something safely, quickly, accurately, and for the lowest cost. Reverse engineering, taking something apart to analyze how it was made or operates, is an important technique used to improve a product or to trouble-shoot a problem in a piece of machinery. Reverse engineering can also be used to analyze a process or system to find ways of making it better. An assembly line is a process used to make a product faster and more efficiently by using uniform parts and following the same steps each time. Industrial engineers are often tasked with designing new assembly line processes, as well as analyzing and optimizing existing processes.

BLAST OFF!

Engineering Fields

- *aerospace engineering*

Engineering Concepts & Skills

- *open-ended problem solving*
- *role of failure*
- *engineering design process*
- *controlled experimentation and testing*
- *modeling*

Estimated time: 40-50 minutes

General Supplies

- *2-3 straws in paper wrappers*
- *masking tape*
- *10 ft. or longer measuring tape*

Supplies Per Family

(one set for **each** family or small group)

- *Blast Off! Design Challenge (Appendix C)*
- *In a gallon-size resealable plastic bag*
  - ☐ *3 sheets of 8 ½" x 11" paper*
  - ☐ *3 standard-size #2 pencils (diameter of pencil needs to be slightly larger than diameter of flexible straw)*
  - ☐ *flexible straw*
  - ☐ *sheet of colored cardstock or construction paper*
  - ☐ *scissors*
  - ☐ *clear tape*
  - ☐ *flexible plastic bottle (all the same kind)—ketchup dispensers with skinny tops work best, but disposable water bottles also work*

Rockets capture our imagination and create excitement. Imagine creating something that weighs millions of pounds, but can fly as fast as 10,000 miles per hour! In this activity, families will design, build, and test paper rockets, discovering what variables affect the speed and distance of their rockets. Like engineers, teams will learn by testing their early designs, and then making modifications and improvements before heading to the launch pad for their final BLAST OFF!

Advance Preparation

- Make double-sided copies of the *Blast Off! Design Challenge* (one per family).

- If the squeeze bottles have very narrow openings, cut a bit off the end so that when the bottle is squeezed there is a quick burst of air. Be sure that the straw still fits snugly over the end.

- Pre-package family supplies into resealable plastic bags for easy distribution. The rolls of tape can be shared between family groups if necessary.

- Remove approximately 2 inches of paper wrapper from one end of a wrapped straw. Use this straw in step 1 below.

ENGINEERING CONNECTION

Aerospace engineering involves teams of scientists and engineers working together. That is because aircraft (including rockets) incorporate many different technologies. Elements of aerospace engineering include **aerodynamics**—the study of the motion of air around an object, **materials science**—what materials should the aircraft be made of or how should fins be attached, and **propulsion**—the energy needed to move the craft through the air.

An important part of the engineering design process is to build and test prototypes, then analyze their performance to improve on the design.

- Set up a "Launch Pad" area. Use masking tape to create a starting line, and then place additional lines with tape every five feet from the starting line to measure distance. A successful rocket may travel up to 30 feet or farther. Make an arrow with tape to indicate the direction of launch. The "Launch Pad" can be as wide as space permits, allowing multiple families to perform their final launches simultaneously. Place a sign next to the "Launch Pad" so that families can easily identify the area.

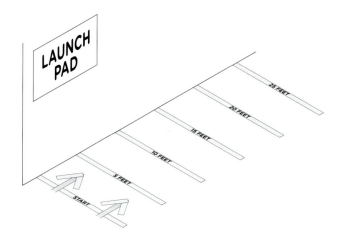

Activity Steps

1. Ask the group to raise their hands if they like rockets. Ask if they have any ideas about how rockets work. Blow a straw wrapper off a straw for everyone to see.

    - How is the wrapper like a rocket?

    - What is it about the wrapper's shape that helps it travel through the air?

    - What made it move?

    - What could be changed to make it fly straighter or farther?

2. Tell families that they are going to work as a team to design a rocket made out of paper and tape. Demonstrate how to make the body of the rocket by rolling a piece of paper around a pencil (using either method shown below) and securing it with tape to make a paper tube. Explain that it is up to each family to decide how much paper and tape to use when making the paper tube. Do they want their design to be heavier or lighter? Longer or shorter?

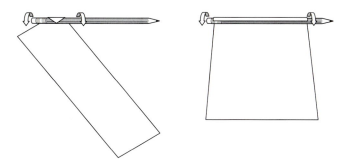

3. Slide the rolled paper "rocket" body off the pencil and put it onto an unwrapped straw (leaving the end of the paper tube open). Blow on the open end of the straw. When nothing happens, ask the group why the rocket body didn't launch. Discuss how the end of the rocket body will need to be sealed so the air doesn't simply pass through it. Point out that because of the failure to launch we were able to discover what was needed to create a better design.

Engineers know that failure is an important tool when testing or refining a new design.

4. Hand out a *Blast Off! Design Challenge* to each family and have them review the challenge.

5. Briefly introduce the engineering design process and encourage families to follow these steps when designing their rockets. See "The World of Engineering" chapter for more information on the engineering design process.

SPECIAL DELIVERY

The engineering used to create rockets impacts our lives in other ways. Everyday we rely on satellites for current weather reports, television transmissions, telephone calls, and to find locations using the Global Positioning System (GPS). How do satellites get into their specific orbits high in space? They are delivered by rockets!

6. Explain that each family will make and test multiple rocket designs to determine which of their designs travels the farthest. Tell groups that every family member should design their own rocket, making each rocket different in some way from the others. Show families the materials they will receive and discuss the changes that can be made to their rockets—changing the length, weight, or shape of the rocket, and adding fins or wings using colored cardstock or construction paper.

7. Explain to families that conducting a **fair test** is important in engineering, so they will want to use the same amount of air for each rocket launch when testing their designs. Show the group how to make a launcher. With tape, secure a flexible straw over the narrow opening of the squeeze bottle or just inside the opening of a plastic water bottle (multiple pieces of tape will be necessary to seal the top of a water bottle).

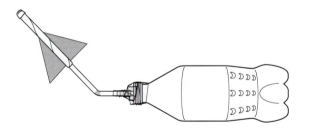

Launcher with water bottle

Launcher with squeeze bottle

8. Let families know that the "Ask" step is an important part of the design process. Asking questions can help define a problem and discover any constraints or requirements, such as what materials can be used. Ask families if they have any questions about the challenge before they begin.

9. Distribute the materials. Ask families to start with 5 minutes of brainstorming possible designs and deciding how each family member's rocket will be different. Family members can use the pencil and paper to draw or list their individual plans.

10. Next, allow families 15-20 minutes to build, test, and modify several rocket designs. Point out that rockets can be tested in the "Launch Pad" area at any time.

11. When families have had time to test and modify their rocket designs, announce to the group that each family should choose one of their rocket designs to be launched as their final BLAST OFF! Encourage families to make any final modifications to the rocket as a team. Allow 5 minutes for final modifications.

12. Gather families behind the starting line of the "Launch Pad" area and have each family launch their final rocket for others to see. Depending on the size of the launch pad area, rockets can be launched all at the same time, or rotating through small groups. Emphasize to families the success that comes with improving their own distance, such as creating a rocket that goes farther than any of their earlier designs, rather than competing with other families.

13. Ask families to share what design elements affected how far their rockets were able to fly (length, weight, fins, wings, etc.). Did they try some ideas that didn't work? How did they solve the problem or decide on a new design? Explain that an important part of the engineering design process is to build and test prototypes, and then analyze their performance to improve on a final design.

Extensions

Encourage families to continue designing and testing rockets at home:

▶ Try changing just one part of your rocket design at a time to see if it changes how your rocket performs.

▶ Try to land your rockets in a specific area or hit a target.

▶ Try making rockets using different materials such as foil, plastic, or thin cardboard.

▶ Try out various squeeze bottles as launchers.

SATURN V ROCKET

When President Kennedy announced in 1961 that the United States would put a man on the moon by the end of the decade, a rocket capable of propelling a manned spacecraft to the moon did not yet exist. The Saturn V was the largest in a series of rockets that were developed to solve the problem of getting humans to the moon. The 3-stage Saturn V was taller than a 36-story building. Inside the rocket was a maze of fuel lines, gauges, pumps, sensors, circuits, and switches—over 3 million parts that needed to work together—which they did! This huge rocket was one of the great achievements of 20th century engineering.

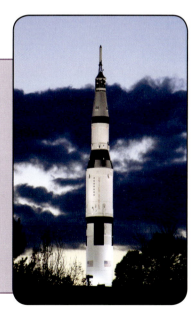

BRAIN SAVER

Engineering Fields
- *biomedical engineering*
- *mechanical engineering*
- *materials engineering*

Engineering Concepts & Skills
- *engineering design process*
- *modeling*
- *properties of materials*

Estimated time: 35-45 minutes

General Supplies
- *plastic tablecloth/dropcloth or newspaper to cover a 5' x 5' square of floor*
- *6' - 8' ladder*
- *10' - 12' tape measure*
- *trash containers*
- *sample helmet(s) (bicycle, football, skiing, etc.)*
- *paper towels for clean up*
- *helmet-making materials*
  - ☐ *variety of hard materials (boxes, plastic containers, egg cartons, plastic cups, etc.)*
  - ☐ *variety of soft materials (bubble wrap, foam, tissue paper, paper towels, newspaper, fabric scraps, felt, etc.)*

Supplies Per Family
(one set for **each** family or small group)
- *Brain Saver Design Challenge (Appendix C)*
- *scissors*
- *ruler*
- *3-4 rubber bands*
- *masking tape (2-3 families can share one roll of tape)*
- *raw egg (additional egg may be needed if retesting design)*
- *resealable plastic sandwich bag*

Head injuries from recreational activities are a leading cause of serious injury, death, and disability in children and adults. Engineers have designed a variety of helmets for different kinds of activities—football, motorcycles, horseback riding, bicycling, skiing/snowboarding, etc. Bicycle helmets, in particular, are 85 percent effective in limiting head and brain injuries from bicycle accidents. In this activity, families design a "helmet" to protect an egg "head" from injury when dropped from a height of 10 feet.

Advance Preparation

- Make copies of the *Brain Saver Design Challenge* (one per family).

- Set up a table or floor space in the room to spread out the helmet-making materials. It helps if the hard and soft materials are in separate areas. *Note: pre-cutting materials such as bubble wrap or fabric into smaller pieces (approximately 5" x 7") will encourage using fewer materials. Rubber bands, tape, and scissors can be passed out to families at their tables.*

- Place each egg in a resealable plastic sandwich bag (remove air and seal, so as not to provide an air cushion). Place the eggs on a separate table from the helmet-making supplies, putting them back in the carton or in a bowl or box to prevent rolling.

- Set up a group testing area by covering a 5' x 5' area of floor with plastic or newspaper and placing the ladder at the edge of this area. Determine what rung of the ladder you will need to stand on in order to hold the "helmets" at a 10-foot height for testing.

Activity Steps

1. With everyone watching, drop a raw egg placed inside a resealable plastic sandwich bag from arm's length to the floor. Ask the following questions:

   - What happened? *(The egg dropped and broke.)*

   - What if that egg was your head, hitting the ground in a fall while riding your bike or rollerblading? *(A head injury would likely occur.)*

   - What should you do to protect your head? *(Wear a helmet.)*

2. Ask the group to think about when it is important to wear a helmet. Quickly read the following list, and ask participants to raise their hands if they think it would be wise to wear a helmet while doing these activities. Recommended responses are in *italics*.

   - Read a book *(no)*

   - Ride a bike *(yes)*

   - Sleep at night *(no)*

   - Downhill ski or snowboard *(yes)*

   - Ride in a car *(no)*

   - Take a walk *(no)*

   - Roller blade *(yes)*

   - Play hockey *(yes)*

   - Play American football *(yes)*

   - Play soccer *(no)*

   - Swimming *(no)*

   - Rock climbing *(yes)*

   - Take a boat ride *(no)*

   - Ride a horse *(yes)*

3. Show the group a real helmet so that they can examine its characteristics. Ask the group what characteristics contribute to good helmet design. Should it be large or small, heavy or light, soft or hard on outside, soft or hard on inside? Tell families that it is not enough for a helmet to just protect your head, it also needs to be comfortable and something that a person is willing to wear. Different aspects of helmet design are usually addressed by a team of engineers from different fields. Biomedical engineers make sure the helmet is comfortable and effectively protects the head during a specific activity. Materials engineers design and select the best materials; and mechanical engineers design how it all fits together and functions effectively.

4. Distribute the *Brain Saver Design Challenge* to each family. Have families review the design challenge and requirements. Let families know that they will have the additional constraint of a limited amount of time in which to design and construct their helmets.

5. Review the engineering design process with the families and encourage them to use it while designing their helmets. See "The World of Engineering" chapter for more information about the engineering design process.

6. Show families the materials available for helmet construction. Ask families to only send one family member at a time to the supply table to gather materials. Remind them that their helmet can only measure 6 inches in any direction, so they should not gather more materials than they need.

7. Let families know that the "Ask" step is an important part of the design process. Asking questions can help them define the problem and discover any constraints or requirements, such as how much time is available or what materials may be used. Ask families if they have any questions about the challenge before they begin.

8. Allow 15 minutes for families to create their helmets. Ask them to start by spending a few minutes imagining and planning their design before they gather materials. Encourage families to continue working on their helmet design until all helmets are ready for testing, adding other "engineered" features to the helmet or even making the helmet more attractive with some decorations.

SMART PROTECTION

One in five high school football players in America suffer concussions—more than 67,000 every year! Engineers are using technology to help prevent this common injury. New helmets are being designed with sensors that send information about each hit to a computer on the sidelines. A special computer program analyzes where the helmet got hit, how hard the hit was, and what direction it came from. Coaches and doctors can then work out a profile for each player and determine the appropriate care for the athlete.

9. When all helmets are ready for testing, gather families around the perimeter of the testing area. One by one, a representative from each family should bring their helmet to the ladder for testing. Either you or an adult assistant should stand on the ladder to drop each helmet. Do not let children climb on the ladder. Tell the group that they will want to be quiet during each drop in order to try to hear if the egg breaks when the helmet hits the ground. After each drop, the family can collect their helmet, but they should not dismantle it until all helmets have been tested. Once all the helmets have been dropped, invite families to open their helmet packages to inspect the condition of their "heads."

10. As a group, compare and discuss the designs and sizes of the different helmets. Which designs were successful? What could have been done differently to better protect their egg "head"? If time allows, have families modify their helmets and re-test as a group again.

11. Have families dispose of broken eggs in the trash container and return unbroken eggs and unused materials to the supply table.

12. Wrap up by explaining that engineers play an important role in designing many different types of safety equipment. Ask the group for examples (seatbelts, electricity outlet covers, shower/bath handholds, non-slip mats, life vests, car seats, etc.).

Extensions

▶ Have families measure and record the size and weight of their helmets and then challenge them to make and test a new helmet that is 10% lighter or smaller. To implement this extension activity, you will need to provide a measuring tape and kitchen-style scale.

ENGINEERING CONNECTION

Biomedical, materials, and mechanical engineers all work together to design effective helmets. A biomedical engineer identifies where the most vulnerable areas of the head are, such as the temple, the forehead, and the back of the head, and how much protection is needed in each area. The materials engineer designs the inner and outer materials that absorb impact, protect those vulnerable areas, and spread the shock of impact throughout the helmet rather than concentrating it in one spot. The mechanical engineer designs systems that secure the helmet on a head and adjust the fit with straps as needed. A successful helmet design must pass a series of performance tests to determine if it will be effective.

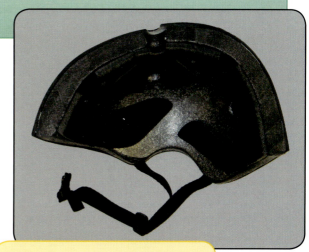

Cross-section of bicycle helmet.

BRIGHT IDEAS

Engineering Fields

- *electrical engineering*

Engineering Concepts & Skills

- *reverse engineering*
- *engineering under constraints*
- *open-ended problem solving*
- *invention/innovation*

Estimated time: 40-50 minutes

General Supplies

- *wire stripper*
- *optional: butcher paper or whiteboard and marker*

Supplies Per Family

(one set for **each** family or small group)

- *Bright Ideas Design Challenge (Appendix C)*
- *2 resealable plastic bags*
- *inexpensive flashlight that uses two D batteries (should be easy to take apart)*
- *2 D Batteries*
- *additional 1.5 volt flashlight bulb*
- *2 three-foot lengths of insulated wire (20 gauge, solid core wire works well)*
- *clothespin*
- *medium-size binder clip*
- *4 pipe cleaners*
- *4 rubber bands*
- *small paperback book*
- *optional: additional items for designing reading light, such as aluminum foil, paper clips, craft sticks, fabric pieces, etc.*
- *masking tape (1 roll may be shared by 2-3 families)*

Once you know how something works, you can "re-engineer" it to make it work more effectively or to do something different with the same materials. In this activity, families will take apart a flashlight, figure out how it works, and transform it into a lightweight reading light.

Advance Preparation

- Try the activity first to become familiar with the procedures.
- Make copies of the *Bright Ideas Design Challenge* (one per family).
- Use wire stripper to remove approximately 1/2 inch of insulation from both ends of each piece of wire.
- Prepare the following three sets of supplies for each family to be distributed during different phases of the activity.

  → Phase One Supplies—place in a resealable plastic bag
    - one D battery
    - extra flashlight bulb
    - one piece of wire with both ends already stripped of insulation

  → Phase Two Supplies
    - flashlight with one D battery inside

  → Phase Three Supplies—place in a resealable plastic bag
    - *Bright Ideas Design Challenge*
    - one piece of wire already stripped of insulation
    - clothespin
    - medium-size binder clip
    - 4 pipe cleaners
    - 4 rubber bands
    - small paperback book
    - optional: additional items for designing reading light

⚙ ENGINEERING CONNECTION

A flashlight is a simple electrical device consisting of a battery, light bulb, reflector, switch, and an electrical circuit. Electrical engineers design electrical circuits to accomplish many different tasks, from powering an iPod to providing electricity and communication systems to an entire city.

Activity Steps

Phase 1. A Simple Circuit

1. Ask families if anyone has ever used a flashlight. How do flashlights work? Point out that flashlights work by using the electricity that is stored in a battery. Show a D battery. Let families know that in this activity they will begin by exploring electricity and how a flashlight works.

2. Explain that there are three phases to the activity. The first is to figure out how to light a bulb with one wire and one battery. The second is to take a flashlight apart and see how it works. The third is to use the flashlight parts to engineer a new product.

3. Show the Phase One supplies—one wire, 1.5V bulb, and one battery. Explain the challenge: **to make the bulb light up using just one wire and a single battery**. Remind families to hold the wire by the insulated portion rather than on the bare wire, because they will be making electricity flow through the wire.

4. Walk around the room, helping families as needed. Be sure to give them time to figure the activity out for themselves.

5. When all or most families have succeeded, ask one or two to explain how they got the bulb to light up. If family members don't mention it, point out that there are two contacts on the bulb: the metal tip at the bottom and the metal side of the bulb base. Each contact needs to be connected to one end of the battery, which can be done with the arrangement shown below. Explain that this makes a complete circuit for the electricity to flow through the bulb and, if one of the contacts is separated, the electricity stops flowing. If there is a white board or large piece of paper available you can draw this single battery and bulb circuit.

A FLASH OF LIGHT

An English inventor named David Missel patented the first flashlight in 1899. It looked similar to modern flashlights. However, early flashlights had batteries that could not provide continuous electrical energy. They produced light in flashes, hence the name "flashlight."

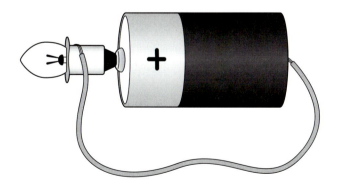

Phase 2. Reverse Engineering

6. Give each family a flashlight containing one D battery. Have the families place the other D battery (from phase one) into the flashlight and turn their flashlights on and off a couple of times to be sure they work.

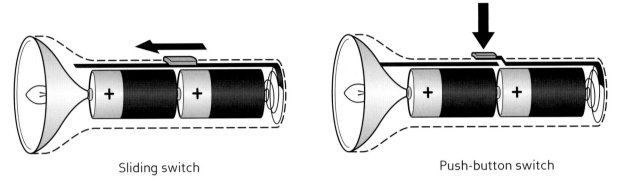

Sliding switch   Push-button switch

7. Explain the challenge: **to take the flashlight apart and see how the switch turns the flashlight on and off.** Walk around and help families as needed, but give them time to discover how the mechanism works. This is a good time to pick up the extra flashlight bulbs from the tables.

8. When families have had a chance to explore their flashlights, ask for explanations on how the flashlights work. Some will probably observe that sliding the switch forward makes a connection between the bottom of one battery and the side of the bulb casing. The base of the bulb is already in contact with the top of one battery, so it is much like the set up in Phase One. If using a push button switch, the circuit is completed as shown in the illustration above. If there is a white board or large paper available you can draw the flashlight switch mechanism as shown above.

9. Explain that taking something apart to see how it works is called **reverse engineering.** Reverse engineering can be used to figure out how to improve a device, or change it to make it perform a different task.

Phase 3. Make a Reading Light

10. Hand out the Phase Three supply bags and distribute masking tape to the tables for sharing.

11. Have families take out the *Bright Ideas Design Challenge* and review the challenge. Point out the "Safety Notes" and also mention that there are many different ways that this problem can be solved.

12. Review the engineering design process with the families and encourage them to use it while designing their reading light. See "The World of Engineering" chapter for more information about the engineering design process.

13. Walk around the room to assist as needed. One possible solution to the challenge, as shown in the illustration, is to use the flashlight body as a battery holder, connecting one wire to the spring (-) and taping the other wire to the top of the battery (+). Then tape the free ends of the wires to the base and side of the light bulb socket. The additional materials can be used to make a book clip, headband, etc.

14. When all or most of the families are finished, ask a few to demonstrate how their reading light works and how it can be positioned for hands-free reading. Chances are good that all the reading lights will be different. Re-emphasize that engineering challenges often have more than one solution.

15. Explain that electrical engineers use their knowledge about how electricity works to design systems that use electricity to get work done or solve a problem. For example, an electrical engineer might design the plan for how electricity will be used to power the streetlights in a city, but an electrician actually installs the wiring and equipment.

Extensions

▶ Challenge families to create a switch system for turning their new reading light on and off.

▶ Suggest that families try taking apart other gadgets or small appliances at home to explore how they operate.

ENLIGHTENING SOLUTION

Incandescent light bulbs may soon become a thing of the past, like typewriters and record players. Electrical engineers are developing alternatives to incandescent lights, such as a compact fluorescent light (CFL), which will help reduce energy consumption. A CFL bulb uses 75% less energy than an incandescent light bulb and lasts 8-12 times longer!

INCANDESCENT LIGHT BULB

COMPACT FLOURESCENT LIGHT BULB (CFL)

However, CFL bulbs contain mercury that is not good for the environment so they need to be disposed of properly. Engineers must often consider a number of trade-offs in order to maximize the benefit to people, the economy, and the environment.

CREATE A CRITTER

Engineering Fields

* *mechanical engineering*

Engineering Concepts & Skills

* *engineering design process*
* *engineering under constraints*
* *reverse engineering*
* *teamwork*

Estimated time: 45-55 minutes

General Supplies

* *5 heavyweight colored file folders*
* *colored markers to share*
* *white cardstock (2 sheets per family)*
* *variety of colored cardstock (2-3 sheets per family)*
* *extra brass fasteners*
* *optional: examples of household mechanisms, such as a hand mixer or pop-up book*

Supplies Per Family

(one set for **each** family or small group)

* *Create a Critter Design Challenge (Appendix C)*
* *critter body—heavyweight colored poster board, approximately 9" x 12" (½ of heavyweight colored file folder will work)*
* *2 sheets of white paper*
* *In a gallon-size resealable bag*
  * ☐ *pencil*
  * ☐ *3 strips of white cardstock (1" x 8 ½")*
  * ☐ *12 brass fasteners*
  * ☐ *clear tape*
  * ☐ *scissors*
  * ☐ *hole-punch*

Mechanical engineers design many different kinds of machines, systems, and mechanisms for getting work done. A mechanism is a device that converts one type of motion to another type of motion, such as when you turn the crank on a hand mixer and the mixers turn to mix the cookie dough. In this activity, families examine some moving mechanisms and try to figure out how the movement is created or controlled. Then families work as a team, using the engineering design process, to design their own mechanisms while creating an imaginary "Critter" with movable parts.

Advance Preparation

* Make copies of the *Create A Critter Design Challenge* (one per family).

* Prepare five different moving mechanism example folders using the directions and illustrations provided.

* Create a sample critter using one or more of the moving mechanisms.

* Cut some of the white cardstock into 1" x 8 ½" strips so that there are 3 strips per family.

* Place the following supplies into a gallon-size resealable plastic bag for easy distribution to each family:

  * pencil
  * 3 strips of white cardstock (1" x 8 ½")
  * 12 brass fasteners
  * clear tape
  * scissors
  * hole-punch

* Place full sheets of white cardstock, colored cardstock, colored markers, and the remaining fasteners on a general supply table accessible to all families.

Activity Steps

1. Ask the group if anyone has used a manual can opener or manual hand mixer. If available, show these or other small household mechanisms and demonstrate how they work. Tell families that these are examples of **mechanisms**. A mechanism converts one type of motion into another type of motion. Ask

families if they have ever looked at a pop-up book. The moving parts of pop-up books are also mechanisms. The motion of turning the book page makes something pop-up out of the book. In some cases, moving or pulling a paper tab makes another element of the page move in a different way.

2. Demonstrate Moving Mechanism #1. Use the term "input" to refer to how the folder is held at the fold and moved left to right. Use the term "output" for the motion of the dangling strips. Point out the difference in the movements of the 2 different sets of colored strips. Ask families what causes the difference in movement between the two types of strips. Then show the inside of the folder and point out that some strips are stapled and attached rigidly to the folder while the other strips have been attached with fasteners, allowing them to move freely.

3. Demonstrate Moving Mechanism #2. Again, using the terms "input" and "output," show the group what happens when the tab slides in and out (input). Ask them what is making the eye move (output)? Show the inside of the folder and how the mechanism works. Point out the guide strips and ask: Why are they needed? Explain that the guide strips help keep the tab in the right location so that it can slide freely from left to right while still being attached to the folder.

4. Distribute a family supply bag and two sheets of white paper to each family. Ask families to take out the cardstock strips and the pencil, but leave the rest of the supplies in the bag.

5. Demonstrate Moving Mechanism #3. Ask families to imagine how things are joined and working inside the mechanism. They can use the cardstock strips, paper, and pencil to help them think about what the mechanism looks like inside. When the folder is opened, point out the differences between the **fixed pivot** and the **moving pivot**, showing that the fixed pivot goes all the way through the folder and the moving pivot does not.

6. Repeat Step 5 with Moving Mechanisms #4 and #5.

MYTHICAL MECHANICAL MASKS

Native Americans of the Pacific Northwest Coast were renowned for their beautiful and elaborate ceremonial masks. Carved from red cedar and painted, these masks often represented animals and mythical creatures. Many of these masks were mechanical. They had moveable parts. Feathers would flap, birds would open their beaks, and eyes would open and close. Using hinges and strings, the performer wearing the mask produced dramatic movements to captivate his audience.

Native American Transformation Mask.

7. Explain that families will be using these mechanisms as they work together on a challenge. Set the moving mechanisms on a table where they can be examined more closely as the families work on their design challenge. Point out that families can examine these mechanisms at the table any time during the activity.

8. Hand out the *Create a Critter Design Challenge* and have families review the challenge. Show families the various supplies and the critter bodies that will be passed out. Remind them that each family will work together to create one critter. Show families the finished sample critter and show how the mechanisms work.

9. Quickly review the engineering design process and encourage the families to use the process while designing their critters. See "The World of Engineering" chapter for more information about the engineering design process.

10. Let families know that the "Ask" step is an important part of the design process. Asking questions can help them define the problem and recall what they know already. Ask families if they have any questions about the challenge before they begin.

11. Ask families to spend five minutes imagining and planning their critters, using the paper and pencil to draw design ideas, before collecting supplies. Encourage them to use at least one idea from each member of the family.

12. After five minutes, hand out one critter body (sheet of heavyweight poster board) to each family. Tell them that they may cut the poster board into the desired shape. Then they can use the white and colored cardstock to make the lever mechanisms and the body parts. Allow 20-25 minutes for families to create their critter as a team.

13. Gather the entire group so that families can briefly share their critters and reveal the mechanisms used to make the moving parts. If some families need more time to complete their critters, explain that they can continue working on their critters at home, adding more parts or features. Alternatively, if some families are finished before others, you can encourage them to be creative in naming and further decorating their critters and inventing behaviors that can be described when introducing their critters to the group.

14. Remind families that what they have just completed is a process that engineers use frequently. They were faced with a challenge and had limited time and materials with which to create a solution. They brainstormed ideas and made a plan. Then they created their critters, testing each mechanism as they were added. When problems came up, they worked together to find solutions.

POP-UP ENGINEERING

Pop-up books are popular and entertaining. The first known book that used movable parts was a book of astronomy in 1306 that used a revolving disc to illustrate its theories. Books with movable parts have been around for centuries, but it was not until the 18th century that the techniques were used in children's books as entertainment. Before that, they were used mainly in scholarly texts to show things such as star movement in the night sky or anatomy of the human body.

Extensions

▶ Encourage families to create more critters at home, using a variety of mechanisms to create motion.

▶ Challenge families to create their own simple pop-up cards or pop-up book using the different mechanisms from this activity.

▶ Suggest that families look for simple machines around their homes. The scissors and hole-punch they have been using are examples of levers. Other examples of familiar mechanisms are listed in the Engineering Connection section below.

Examples of Critters

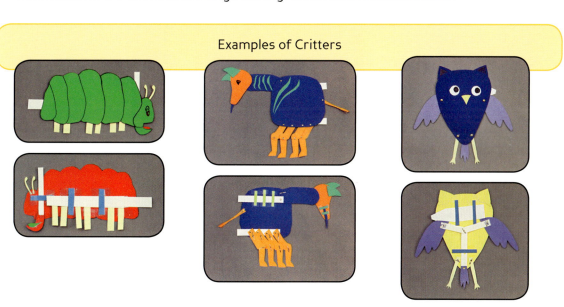

☀ ENGINEERING CONNECTION

Mechanical engineers design mechanisms and machines. There are six different types of simple machines—levers, pulleys, inclined planes, wedges, screws, and the wheel and axle. These simple machines change the direction or size of a force and make it easier to "do work." Here are some familiar examples of simple machines:

Gears—clock, bicycle gears, hand drill, hand mixer, manual can opener
Lever—scissors, door on hinges, seesaw, claw hammer, bottle opener, nail clippers, tweezers, nutcracker, wrench, fishing rod, stapler, crowbar
Inclined Plane—staircase, ramp, bottom of a bathtub
Pulley—flagpole, crane, window blinds
Wedge—axe, zipper, knife, log splitter, doorstop, shovel
Screw—bolt, spiral staircase, jar lid
Wheel & Axle—bicycle, wheelbarrow, car tires

Mechanical engineers use their understanding of simple machines and "how things work" to create a wide variety of machines and mechanisms to solve problems or meet specific needs.

Directions for making Moving Mechanisms

1. Using the colored file folders, white cardstock strips, and colored cardstock, assemble each mechanism inside a file folder according to the photographs and instructions below.

2. Punch necessary holes with a hole-punch. To punch holes in the middle of the file folder (where the hole-punch cannot reach), use a sharp pencil.

3. Fixed pivots are circled in the photos. The fasteners for fixed pivots go through the folder. Moving (or floating) pivots are NOT attached to the folder and should only go through the moving parts allowing the pivot to move freely. Sometimes, to improve the movement, a fastener will need to be loosened, so that it is not fitting too tight to the folder. Once assembled, circle the fixed pivots so that they can be easily pointed out to the families.

4. Cut small strips of colored cardstock to use as guides or tracks. Tape these on as shown to keep the white cardstock strips in place. Use contrasting colors for the guide strips to help families see the parts of each mechanism.

5. Number the folders 1 through 5 so the mechanisms can be shown in that order.

Moving Mechanism #1—Staple 2 strips to the back of the file folder as shown. Use a hole-punch and fasteners to attach 4 other strips in a contrasting color as shown. Close the folder. Holding the folder at the fold, shake left to right to make the 4 strips move.

Outside   Inside

Moving Mechanism #2—Cut a 3" x 12" strip of poster board. Cut an opening in the front of the file folder. Using guide strips, secure the poster board strip behind the opening, leaving a tab exposed to pull. Use a marker to make an oval in the center of the "eye." Close the folder. Move the tab slightly left to right to make the eye move.

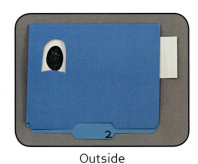

Outside

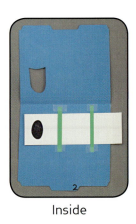

Inside

Moving Mechanism #3—Cut a hand or mitten shape from cardstock and attach it to the end of the lever strip with tape. Create the mechanism as shown. The fastener for the fixed pivot should go through both the lever arm and the folder and is circled. The moving pivot only attaches the two moving strips together.

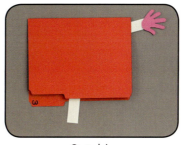
Outside

Inside

Moving Mechanism #4—Cut "paw" shapes from cardstock, color with marker, and attach with tape. Create the mechanism as shown, using two fixed pivots (circled) and one moving pivot. For the fixed pivots, the fasteners should go through the paw-arm and the folder. For the moving pivot, the fastener should go through three moving strips only.

Outside

Inside

Moving Mechanism #5—Cut out a turtle's head, or something to go in and out from the side of the folder, and attach to the lever strip with tape. This mechanism needs a triangle shape with one fixed pivot (circled) and two moving pivots. For the fixed pivot, the fastener should go through the triangle and the folder. For the moving pivots, the fastener should go through the triangle and the moving strip only.

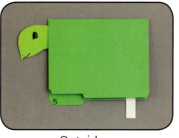
Outside

Inside

ENGINEERING ALL AROUND

Engineering Fields

- *general engineering*

Engineering Concepts & Skills

- *communication*
- *teamwork*
- *sustainability*
- *invention/innovation*

Estimated time: 25-35 minutes

Supplies Per Family

(one set for **each** family or small group)

- *Large Game Board, Side A and Side B (Appendix D)*
- *file folder*
- *set of Engineering All Around Game Cards (Appendix D)*
- *5 different small objects for game tokens (one for each player)—such as different colored beads, buttons, or paperclips. Have extra game pieces available for larger groups.*
- *small, resealable plastic bag*

Every human-made object around us has been engineered for a particular purpose. The Engineering All Around game is a fun way for families to explore engineered objects all around them and to learn just how important engineering is in our everyday lives.

Advance Preparation

- Copy the two halves of the *Large Game Board* and glue or tape them onto the inside of a file folder to create a full game board when opened up flat.

- Copy the two-sided *Engineering All Around Game Cards* on colored card stock and cut apart. *Note: if the game board and cards will be used multiple times, you may want to laminate them so they last longer.*

- Place a set of cards and game tokens into a small resealable plastic bag for easy distribution.

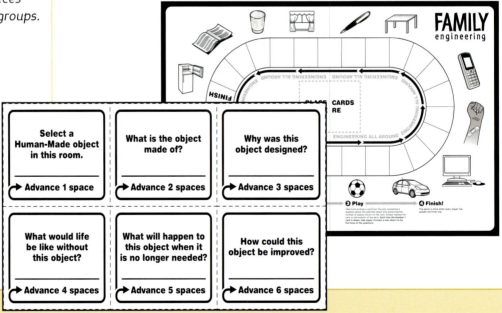

Engineering All Around Game Board and *Game Cards* are available in Appendix D.

Activity Steps

1. Tell families that most of the things we see around us, even things as simple as a cup or pencil, were designed by engineers to solve a problem or to meet a need. For example, point out a chair in the room. Ask families why we have chairs (sit off the floor). Next, ask someone to describe what the chair is made of (metal, wood, plastic, or a combination). Finally, ask families to take 2-3 minutes at their tables to discuss how this particular chair could be improved. Have a few tables report their ideas to the whole group.

2. Explain to families that they will be playing the Engineering All Around game. To play, they will select an engineered object in the room with them, then answer questions about that object. The questions will be similar in nature to the ones they just answered about the chair.

3. Pass out the game boards and bags containing cards and game tokens. Explain that this is a cooperative game rather than a competitive game. The game is over when **everyone** gets past the finish line. So if a player is stuck on a question, other family members may help them out with hints.

4. Ask the families to open their game boards and put them on the table so that all family members can reach the board. Ask family members to read the instructions on the game board as you read them aloud. Answer any questions families have about the instructions.

5. Allow families 20 minutes to play as many rounds of the game as possible.

6. After 20 minutes have passed, ask 1-2 families to share the object that they found to be the most interesting.

A GROWING CONCERN

It is important to think about what happens to a product after it is no longer needed or wanted. Between 1980 and 2008, the generation of solid waste from cities has grown by over 60%—to nearly 250 million tons per year! (Source: U.S. Environmental Protection Agency).

ENGINEERING CONNECTION

Engineering is all around us. Every human-made object or technology has been designed or engineered to solve a particular problem or to meet a need. Civil engineers design buildings and highways. Materials engineers create new kinds of materials for clothing, cars, bandages, or buildings. Manufacturing engineers design the processes for mass-producing items. Chemical engineers design processed, pre-packaged food.

Extensions

▶ The game can be made more challenging by choosing processes and systems rather than products. For example, processes might include vacuum cleaning, brushing teeth, taking a bath or shower, while systems might include a water treatment and delivery system or the electrical system.

ENGINEERING CHARADES

This activity uses the well-loved game of charades to introduce families to a variety of engineering fields and careers and helps them to recognize that many familiar objects we use every day are, in fact, engineered. Families have fun taking turns acting out different engineered products while other families guess which product is being acted out.

Engineering Fields

- *general engineering*

Engineering Concepts & Skills

- *communication*
- *modeling*
- *teamwork*

Estimated time: 20-30 minutes

General Supplies

- *Engineering Charades Cards—set of 12 for every 2-3 families (Appendix D)*
- *letter-sized envelopes to hold each set of cards*
- *Engineering Charades Game Instructions for each family (Appendix C)*

Advance Preparation

- Make enough color copies of the *Engineering Charades Cards* to have a set of 12 cards for each group of 2-3 families. Cut them into individual cards and place one set of 12 cards in each envelope. *Note: if the game cards will be used multiple times, you may want to laminate them so they last longer.*

- Make copies of the *Engineering Charades Game Instructions* handout (one per family).

MECHANICAL ENGINEER

Conveyor Belt

MATERIALS ENGINEER

Waterproof Fabric

CIVIL ENGINEER

Bridge

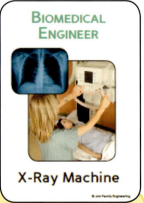

BIOMEDICAL ENGINEER

X-Ray Machine

Sample of the *Engineering Charades* cards available in Appendix D.

Activity Steps

1. Ask families to raise their hands if they have played charades before. Explain that charades is a guessing game where players act out an object, word, or phrase while others try to guess what it is. Engineering Charades is a game where families will work as a team to act out a variety of "products" designed by engineers. Show families one of the game cards showing an engineered object and labeled with an engineering field. Explain that most products are designed by a team of engineers and that these cards list only the main field of engineering that would be involved in the design of each object.

2. Working with 1-2 other event volunteers, demonstrate how to play the game. Use an example that is not on the cards, such as a house designed by a civil engineer or a cell phone designed by a computer engineer. Begin by announcing out loud the engineering field; then, working as a team but without speaking, act out the engineered object and have the families guess what object you are acting out. Remind families that they will be working as a team to act out and guess each engineered object. Encourage them to involve all family members when acting out the charade.

3. Divide everyone into groups of 2-3 families (total of 10-12 children and adults combined).

4. Pass out the *Engineering Charades Game Instructions* to each family and review.

5. Distribute an envelope of Engineering Charades cards to each group of 2-3 families.

6. Allow 15-20 minutes for families to play the game. Collect the envelopes and cards.

7. Wrap up by pointing out that engineers design just about everything we come in contact with or use in our everyday lives. This means that there need to be a lot of engineers in the world, working in a variety of different fields of engineering.

ENGINEERING CONNECTION

Products and structures don't just "happen." They are imagined, designed, and engineered by people. There are many different fields of engineering. Engineers design products that we use everyday.

FIVE POINTS TRAFFIC JAM

A transportation engineer can save lives by designing safer intersections. In this activity, families imagine they are transportation engineers and take on the challenge of designing an intersection that will be safer for car passengers, bicyclists, and pedestrians.

Engineering Fields

- *civil engineering*
- *systems engineering*

Engineering Concepts & Skills

- *engineering design process*
- *modeling*
- *spatial visualization*
- *teamwork*
- *optimization/tradeoffs*

Estimated time: 30-40 minutes

General Supplies

- *tape*

Supplies Per Family

(one set for **each** family or small group)

- *Five Points Traffic Jam Design Challenge (Appendix C)*
- *2 Small Intersection Maps (Appendix D)*
- *Large Intersection Map, Sides A and B (Appendix D)*
- *paperclip*
- *In a resealable plastic bag*
  - ☐ *2 pencils*
  - ☐ *3-4 fine-point washable colored markers*
  - ☐ *scissors*
  - ☐ *glue stick or tape*

Advance Preparation

- Make copies of the *Five Points Intersection Design Challenge* (1 per family).

- Make copies of the *Small Intersection Map* on 8 ½" x 11" paper (2 per family).

- Assemble one *Large Intersection Map* per family. Copy both side A and side B on 8 ½" x 11" paper and tape them together.

- Make color copies of the *Traffic Management Tools* on cardstock and cut in half (one half page per family).

- Paperclip the following handouts together for easy distribution to each family: *Five Points Traffic Jam Design Challenge* (on top), *2 Small Intersection Maps*, *Large Intersection Map*, and *Traffic Management Tools*.

- Place 2 pencils, 3-4 fine-point colored markers, scissors, and a glue stick or tape into a resealable plastic bag for easy distribution to each family.

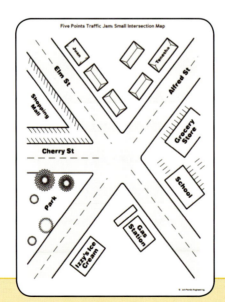

Intersection Maps and *Traffic Management Tools* are available in Appendix D.

Activity Steps

1. Ask families what they think a transportation engineer does (see Engineering Connection).

2. Introduce the activity by explaining that transportation engineers can save lives by designing safer intersections. Ask families if they have experienced a dangerous road intersection, either as drivers, passengers, bike riders, or walkers. Ask what kinds of things make an intersection dangerous *(cars driving too fast, no signs, no crosswalks, etc.)*. Ask the group to suggest ways to make intersections and streets safer *(speed limits, stop signs, traffic signal, crosswalks, bike lanes, sidewalks, traffic circles, pedestrian bridges, etc.)*.

3. Pass out the packet of handouts and the resealable plastic bag of supplies. Have families open the *Large Intersection Map* and explain that this is an intersection where five different roads come together. Have families look at the map and discuss where people might get injured or cars might collide.

4. Tell families to imagine they are a team of transportation engineers hired to make this intersection safer for car passengers, bicyclists, and pedestrians. Review the design challenge and show the *Traffic Management Tools* sheet that families have in their packets. Point out that they can use the small maps and pencils for planning and brainstorming their ideas and discussing safe routes for Tenesha, Juan, and Samantha. Then they can create their final design on the larger map, using colored markers and the traffic management tools. Tell families that they are not limited to just the "cut out" tools. They can add crosswalks, bridges, roundabouts, or even move a roadway in order to create a safer intersection.

TIME TO WALK

The first traffic light was installed in London in 1868 to control the movement of pedestrians and horse-drawn buggies. It was a lantern with red and green signals and it had to be operated manually by a policeman! Some early traffic signals also had a 'gong' so that drivers would see and HEAR the traffic signal. Today there are "Walk" signals at some crosswalks that have timers, telling the pedestrian how many seconds they have until the light changes.

✳ ENGINEERING CONNECTION

Transportation engineering is a part of civil engineering. Transportation engineers are concerned with the safe and efficient movement of people and goods. They design highways, railroads, bike paths, airports, city streets, shipping ports, and subway systems to help people travel by car, bike, train, plane, bus, and even on foot.

5. Review the engineering design process with families and encourage them to use it while designing a safe intersection. See "The World of Engineering" chapter for more information about the engineering design process.

6. Allow families 10 minutes of planning and discussion time and then tell them that they will have 15 minutes more to create their plan on the *Large Intersection Map*.

7. After 15 minutes, ask families to share their plans with a family next to them, allowing time for both families to present their ideas.

8. Ask the families what traffic management tools were most commonly used. Which ideas would likely cost the most money? The least amount of money? Explain that engineers often have to make trade-offs in a design in order to balance different needs, including things like costs, available resources, and the impact on the environment.

Extensions

▶ Have families draw a real intersection in their hometown, then redesign it to make it safer using a variety of traffic management tools.

One of the world's worst traffic jams was called the "Great Chinese Gridlock of 2010" by the New York Times. It lasted for 11 days in August of 2010 in the Hebei Province of China. Traffic congestion stretched for over 62 miles and some drivers spent 5 days getting through the traffic jam!

TOUGH TRAFFIC

Imagine having no hands. Holding a pencil, tying your shoes, or simply picking something up would be very challenging. Biomedical engineers help solve this problem by designing prosthetic limbs—artificial devices that replace a missing body part. By doing so they can literally "give someone a hand"! In this activity, families will design a simple prosthetic hand—a device that can be used to pick up objects.

Advance Preparation

- Make copies of the *Give Me a Hand Design Challenge* (one per family).

- Place the following materials into a resealable plastic bag for easy distribution to each family—6 wooden craft sticks, 8 rubber bands, 3 paperclips, 2 index cards, a plastic spoon, and a plastic fork.

- Place the test objects on a centrally located table—cotton balls, erasers, pencils, a container of marbles, paper, and plastic cups. Place a sign on or near the table to identify it as the "Testing Table."

Activity Steps

1. Ask the group to imagine having only one hand or no hands at all. Have everyone make fists with their hands and try to pick something up. Next, ask them to consider trying to tie a shoe, button a shirt, eat dinner, or write with a pencil.

2. Explain to families that a biomedical engineer can help solve this particular problem by designing a prosthetic hand. A prosthesis is an artificial device that replaces a missing body part, and they have come a long way from when they were just a motionless attachment to take the place of a missing hand or leg—like Captain Hook's hook in Peter Pan. These days there are competitive runners who wear special prosthetic legs and prosthetic hands where the fingers move when triggered by electrical signals from arm muscles.

Engineering Fields

- *biomedical engineering*
- *mechanical engineering*

Engineering Concepts & Skills

- *engineering design process*
- *engineering under constraints*
- *properties of materials*

Estimated time: 30-40 minutes

General Supplies

- *test objects (one of each for every 3-4 families)*
  - ☐ *cotton balls*
  - ☐ *erasers*
  - ☐ *pencils*
  - ☐ *marbles (in a container)*
  - ☐ *8 oz. plastic cups*
  - ☐ *sheets of paper*

Supplies Per Family
(one set for **each** family or small group)

- *Give Me a Hand Design Challenge (Appendix C)*
- *pencil*
- *sheet of white paper*
- *masking tape (one roll can be shared with multiple families)*
- *in a resealable plastic bag:*
  - ☐ *6 wooden craft sticks*
  - ☐ *8 rubber bands of various sizes*
  - ☐ *3 paperclips*
  - ☐ *2 index cards*
  - ☐ *plastic spoon*
  - ☐ *plastic fork*

3. Explain that today each family will work as a team of biomedical engineers to create a simple prosthetic hand—a grabber device that can pick up a variety of objects.

4. Distribute the *Give Me a Hand Design Challenge*, a sheet of paper, and a pencil to each family and review with them the design challenge and requirements. Show families the materials that will be distributed later.

5. For this activity, families should imagine that their thumbs and fingers will act as the electrical signals from the arm muscles that make a prosthetic hand move.

6. Quickly review the engineering design process with families and encourage them to use it while designing their grabber device. See "The World of Engineering" chapter for more information about the engineering design process.

7. Let families know that the "Ask" step is an important part of the design process. Asking questions can help them define the problem and consider any constraints or requirements, such as what materials can be used to build a device or what a device needs to be able to do. Ask families if they have any questions about the challenge before they begin.

8. Have families begin by spending a few minutes brainstorming ideas and planning their designs. They can use the paper and pencil for drawing their plans.

9. After a few minutes, pass out a bag of materials to each family. Place a roll of masking tape to share on each table. Let families know that they will have 15-20 minutes to build and test their grabber devices. Remind families to keep testing their designs and making improvements as they go along. Show them the "Testing Table" where they can test their grabber devices. Point out that if they succeed at picking up the first four objects in the challenge, they can accept another challenge by trying to pick up other objects, such as those listed in the "Ready for More?" section.

10. When the grabber devices are completed, allow time for families to share their designs with other families, pointing out approaches that worked well, and discussing some of the challenges they faced and adjustments they made as a result of their testing.

ENGINEERING CONNECTION

Biomedical engineers enhance the health and well being of people by engineering solutions to complex medical problems. Biomedical engineers design prosthetics, diagnostic equipment, examination instruments, software, and medical devices to help physicians, nurses, therapists, and technicians enhance the quality of medical care and life for millions of people around the world.

11. Share with families that biomedical engineers often work on products that are meant to enhance the quality of life and medical care for patients. They help doctors, nurses, and others in the medical field to care for their patients by designing instruments, tools, and medical devices. In addition to prosthetic limbs, a biomedical engineer may also design imaging devices, such as an X-ray or MRI machine, systems for delivering treatments and medicines (such as an IV or dialysis machine), tools for diagnosing disease or injury, and treatments or procedures to help patients recover from an injury or function as normally a possible following a debilitating accident or disease.

AMAZING INVENTION

The i-LIMB™ Hand, developed by Touch Bionics in 2007, was the first prosthetic hand that allowed all five fingers to move individually. The fingers are controlled by tiny electrical signals generated in the arm muscles. Mechanical and biomedical engineers used ground-breaking techniques to design this prosthetic device, making it operate like a real hand, while also being lightweight, durable, and appealing to a patient because it looks like a real hand. The company also developed LIVINGSKIN™, a silicone material that closely mimics the appearance of a patient's own natural skin while moving and flexing in the same way that human skin does.

HOT CHOCOLATE MACHINE

Engineering Fields

* *chemical engineering*
* *mechanical engineering*

Engineering Concepts & Skills

* *engineering design process*
* *open-ended problem solving*
* *engineering under constraints*
* *modeling*

Estimated time: 40-50 minutes

General Supplies

* *measuring spoon (teaspoon) or small plastic spoon*
* *source for warm water*
* *container/pitcher for dispensing water*
* *6 additional 5 oz. cups*
* *bucket or plastic bin*

Supplies Per Family

(one set for **each** family or small group)
* *Hot Chocolate Machine Design Challenge (Appendix C)*
* *sheet of white paper*
* *pencil*
* *22 paper cups (5 oz.)*
* *2 paper or plastic cups (8 oz.)*
* *short pushpin with plastic head*
* *plastic tray or shallow pan (with flat bottom)*
* *1 cup of cool water*
* *½ cup of warm water*
* *2 tsp. hot chocolate powder*
* *2 tsp. powdered milk*
* *2 paper towels*
* *masking tape (one roll can be shared by 2-3 families)*
* *optional: additional chocolate and milk powder if allowing time for design improvements and a second test.*

Ever wonder how hot chocolate, soda, or milkshake machines work? First, food chemists experiment with preparing these foods in a lab. Next, chemical engineers design the process for mixing large quantities of the food product and mechanical engineers design the machines that can mix and deliver the final product. In this activity, families will work together to design, construct, and test a machine that can mix up a cup of hot chocolate.

Advance Preparation

* Make copies of the *Hot Chocolate Machine Design Challenge* (one per family).

* Place twenty empty 5 oz. paper cups, one 8 oz. cup filled with cool water, a pushpin, and 2 paper towels onto a tray. Make one tray for each family.

* Put 2 teaspoons of chocolate powder in a 5 oz. cup and 2 teaspoons of milk powder in another 5 oz. cup (one of each per family) and place these on a supply table for families to gather during the activity. *Note: If doing a second round of testing, keep remaining chocolate and milk powder available for refilling the cups.*

* Identify a source of warm water (warm tap water works fine). While families are working on their designs, fill one 8 oz. cup **half full** of warm water for each family. *Note: if doing a second round of testing, refill the cups half full of warm water just before conducting the second test.*

* Gather six 5 oz. cups (one with a hole in the bottom or side) and a cup of water for demonstrating how cups can be stacked and water can flow from one cup to another.

* Place the bucket or plastic bin on the supply table for collecting liquid waste.

* Using the activity steps below, try making a hot chocolate machine yourself so that you will be able to answer questions and help guide families through the challenge.

Activity Steps

1. Ask the group if they have ever gotten a cup of hot chocolate (or other kind of drink such as coffee or soda) from a machine. When you press the button, out comes a cup of hot chocolate! How does the machine make the hot chocolate? Is it magic? No, it's engineering!

2. Remind families that a hot chocolate machine looks like just a box on the outside, but the real work happens inside the box. Chemical engineers determine the right mixture of ingredients and how the ingredients need to be mixed. Then mechanical engineers design the machine that does the mixing and dispenses the drink into a cup.

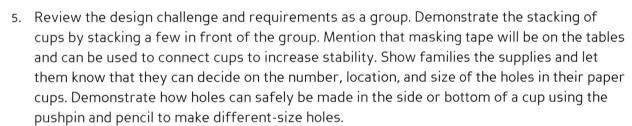

3. Explain to families that they will be designing a machine to mix a cup of hot chocolate. These machines will not use electricity to do the work. Instead, they will use gravity. Demonstrate how gravity drains the water out of a paper cup with a hole in the bottom or side.

4. Hand out the *Hot Chocolate Machine Design Challenge*, a piece of paper, and a pencil to each family.

5. Review the design challenge and requirements as a group. Demonstrate the stacking of cups by stacking a few in front of the group. Mention that masking tape will be on the tables and can be used to connect cups to increase stability. Show families the supplies and let them know that they can decide on the number, location, and size of the holes in their paper cups. Demonstrate how holes can safely be made in the side or bottom of a cup using the pushpin and pencil to make different-size holes.

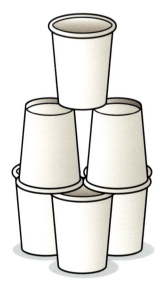

6. Explain to the families that they will design and test a gravity flow "machine" based on stacking paper cups so that powders and liquids mix as the water flows due to gravity. Families will be given one cup of cold water for initially testing the water flow in their machines. The powders and warm water will be distributed once all the machines have been tested for flow and are ready for testing the mixing of all materials.

7. Quickly review the engineering design process with the families and encourage them to use it while designing their machine. See "The World of Engineering" chapter for more information about the engineering design process.

8. Let families know that the "Ask" step is an important part of the design process. Asking questions can help them to define the problem and consider any constraints or requirements, such as what their hot chocolate machine needs to do or what materials can be used. Ask families if they have any questions about the challenge before they begin.

Cocoa beans have been grown as a crop in the Amazon basin since 1500-400 BCE. The Mayans and Aztecs used them for both food and money. Chocolate has been used as a drink since this time. However, Europeans did not know about the popular drink from Central and South America until the Spanish introduced it to them in the 16th century.

Cocoa bean pods on tree.

9. Give families 5 minutes to brainstorm ideas and plan their machine by drawing possible cup arrangements with the pencil and paper. Remind them to consider where they might put the holes, how the water will flow through the machine, and how the liquids and powders will be mixed.

10. After the planning phase, distribute masking tape to the tables and have families pick up a tray of supplies. Allow families 15-20 minutes to build and test the water flow through their machines using only the cool water. Remind families to build and test their machine on the tray to catch any spills.

11. While families are building and testing their machines, prepare an 8 oz. cup **half full** of warm water for each family and place the cups on the supply table.

12. When the machines have been tested for flow, have one member from each family collect the two powders and another member collect the cup of warm water. Tell families to load their machines by placing the powders in the appropriate cups, but DO NOT pour the warm water until you give the signal. Once all machines are loaded and ready to go, signal for one family member to carefully pour the warm water into their machine. Encourage families to observe their machines in action.

13. Ask families to discuss with each other how their hot chocolate machines worked. Have a few families share some successful design strategies or share parts of their designs that did not work. Ask families to consider how their designs could be improved.

A PENNY FOR SOME WATER?

The first vending machine was invented by Hero of Alexandria, a Greek mathematician and engineer who was born in 10 AD. The machine dispensed a set amount of water when a coin was put in a slot. The coin fell into a pan and its weight lifted a small cork to release water. As the pan tipped, the coin would fall off and the pan and cork would snap back into place to shut off the flow of water.

14. Optional: If time allows, inform families that they will have a chance to improve their designs. Let them know that they can drink their first cup of hot chocolate or dispose of it in the waste bucket, then wipe out their cups with the paper towels. Allow 5-10 minutes to modify the machines. While families are modifying their designs, refill cups of powder (two teaspoons in each) and the half full cup of warm water for each family. Repeat the testing process and discuss whether their design modifications made a difference in the way their hot chocolate machines worked.

Example of Hot Chocolate Machines: Be sure **not** to share this photo with families before they have a chance to try the challenge on their own.

Extensions

▶ Challenge families to try this activity at home and create new requirements for their machines, such as using fewer cups, delivering hot chocolate in less time, adding a third powder or marshmallows, etc.

ENGINEERING CONNECTION

Chemical engineers work with raw materials to produce useful products and processes. Many chemical engineers work in the food processing industry. They are often asked to design procedures that deliver food materials and/or liquids in a specific quantity over a specific amount of time. They are also often asked to try to use the least amount of energy and materials, and produce the least amount of waste. When designing something like a hot chocolate machine, chemical engineers team up with mechanical engineers to create an effective machine.

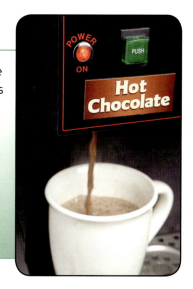

LAUNCHER

Engineering Fields

- *mechanical engineering*
- *aerospace engineering*

Engineering Concepts & Skills

- *engineering design process*
- *engineering under constraints*
- *open-ended problem solving*

Estimated time: 30-40 minutes

General Supplies

- *masking tape*
- *tape measure*
- *marker*
- *replacement materials—craft sticks, rubber bands, plastic spoons, cotton balls*

Supplies Per Family

(one set for **each** family or small group)

- *Launcher Design Challenge (Appendix C)*
- *Real Life Launchers photo page (Appendix D)*
- *8 ½" x 11" sheet of paper*
- *pencil*
- *masking tape (1 roll can be shared among 2-3 families)*
- *in a resealable plastic bag:*
  - ☐ *12 craft sticks*
  - ☐ *6 rubber bands*
  - ☐ *plastic spoon*
  - ☐ *cotton ball*

Have you ever used a ball thrower or a diving board? These are examples of real-life "launchers"—machines that propel an object. Launchers work by storing potential energy and then releasing it as kinetic energy (energy of motion) which is used to propel an object. In this activity, families will design their own "launchers."

Advance Preparation

- Make copies of the *Launcher Design Challenge* (one per family).

- Make copies of the *Real Life Launchers* photo page (one per family).

- Place the following materials into a resealable plastic bag (one bag per family): 12 craft sticks, 6 rubber bands, plastic spoon, and cotton ball.

- Set up the "Launcher Testing Grounds" by using masking tape and a marker to create a starting line. Then, mark every two feet with masking tape to allow for easy measurement of distance during launch testing. It may be helpful to post a sign for the "Launcher Testing Grounds" to help families locate this area.

- *Event Tip:* Have extra craft sticks, plastic spoons, rubber bands, and cotton balls available in case materials are lost or damaged during the activity. If materials are replaced, be sure families limit their materials to the original number of items in the bag.

Activity Steps

1. Introduce the group to the idea of a "launcher"—a machine that stores energy and can be used to propel an object. Demonstrate how your finger and a rubber band can be used as a launcher. The stretched rubber band is an example of stored potential energy. Explain that when the rubber band is let go, the potential energy is released and becomes kinetic energy—the energy of motion. Without pointing the rubber band at another person, send it flying across the room. Ask families if they can think of other examples of launchers (e.g. baseball pitching machine, diving board, tennis ball launchers, pop-up toaster, etc.). Explain that mechanical engineers design lots of different kinds of machines. A machine is an object that uses or changes energy to do work or complete a task. A launcher's task is to propel an object.

2. Hand out the *Launcher Design Challenge* and the *Real Life Launchers* photo page, a blank sheet of paper, and a pencil to each family and have them review the design challenge and requirements. Show families the materials that will be distributed to each team, including the masking tape that will be available at the tables.

3. Review the engineering design process and encourage families to use it while designing their launchers. See "The World of Engineering" chapter for more information about the engineering design process.

4. Let families know that the "Ask" step is an important part of the design process. Asking questions can help families to define the problem and consider any constraints or requirements, such as what their launcher needs to do or what materials can be used. Ask families if they have any questions about the challenge before they begin.

5. Explain that families will have 5 minutes to discuss ideas and possible solutions to the challenge. They can use paper and pencils for planning and drawing their ideas.

FLYING PUMPKINS

The town of Bridgeville, Delaware attracts thousands of people each year when it hosts their annual Punkin' Chunkin' contest. Participants gather to launch pumpkins across a field using homemade launchers!

6. After 5 minutes, pass out a bag of materials to each family, letting them know that they will have 15 minutes to build their launchers. Remind families to keep testing their designs and making improvements as they go. Show them the "Launcher Testing Grounds" and ask that they test their launchers in this area only.

7. When families are finished testing and improving their launcher designs, invite them to gather at the "Launcher Testing Grounds" for a group launch. Depending on the size of the testing grounds, all teams can launch simultaneously or small groups can take turns launching while the other families observe.

8. When the final launches are complete, discuss, as a group, the different launcher designs.

    •   *Which launchers were the most effective? Why?*

    •   *Did the most effective launchers have any design elements in common?*

    •   *Ask families to share some ideas that didn't work.*

9. Explain to the group that engineers use their knowledge of science and math to help them design new products and create new inventions. Understanding potential and kinetic energy helped the families to design effective launchers.

PLAY BALL

Jai alai is a ball game that originated in Spain but is also played in Mexico, U.S., Brazil, Philippines, Italy, Indonesia, China, and Egypt. It is played in a three-walled court with a hard rubber ball that is caught and thrown with a cesta, a long, curved wicker scoop strapped to one arm. When attached to a player's arm, the cesta becomes a type of launcher, capable of propelling the 125-gram ball up to 180 miles per hour.

Extensions

▶ Challenge families to design a launcher that will work with the greatest accuracy. For example, design a launcher that can shoot a cotton ball into a container at a distance of 5, 10, or 15 feet.

ENGINEERING CONNECTION

There are a variety of real-life uses for launch systems— launching aircraft from ships, launching tennis balls or baseballs for practice, launching "divers" into a swimming pool, or even throwing a ball for a pet dog. Engineers have taken the basic concepts about the properties of motion and used them to launch a range of objects in unique ways. For engineers, knowledge and understanding of science are valuable tools for doing their work.

MINING FOR CHOCOLATE

Engineering Fields

- *mining engineering*
- *environmental engineering*
- *geological engineering*

Engineering Concepts & Skills

- *modeling*
- *sustainability*
- *systems*

Estimated time: 35-45 minutes

General Supplies

- *marker*

Supplies Per Family

(one set for **each** family or small group)

- *Mining for Chocolate Design Challenge (Appendix C)*
- *hard chocolate chip cookie (Chips Ahoy!® work well)*
- *soft chocolate chip cookie (Chewy Chips Ahoy!® work well)*
- *In a quart-size resealable plastic bag:*
  - ☐ *pencil*
  - ☐ *2 paper towels*
  - ☐ *2 small white paper plates*
  - ☐ *2 toothpicks*
  - ☐ *2 paperclips*
  - ☐ *2 craft sticks*
  - ☐ *plastic spoon*

Mining engineers try to extract the largest amount of mineral resources using the least amount of money but providing the greatest amount of safety for the miners. At the same time, engineers must minimize the amount of damage that they do to the environment. This is quite a challenge! In this activity, families will create a process for extracting the valuable resource of chocolate chips from a cookie while dealing with constraints on time, materials, and "environmental" impact.

Advance Preparation

- Make double-sided copies of *Mining for Chocolate Design Challenge* (one per family).

- Use a marker to label one paper plate "soft" and the other paper plate "hard" for each family. Alternatively, you can have families do this themselves with pencils after handing out supplies.

- Place family supplies into the plastic resealable bags for easy distribution. Cookies should be handed out separately. *Note: If soft cookies are stored in the same container with hard cookies, the hard cookies may get soft.*

- *Event Tips:* Have extra supplies and cookies available for large family groups that may need more than one cookie in order to allow everyone to be engaged in the activity. Also, to alleviate the temptation for eating the materials (cookies), you may want to provide additional cookies as a snack at the end of the activity.

Activity Steps

1. Ask the group to think of things that humans extract, or mine, from the earth (coal, oil, gas, diamonds, iron, gold). Ask why they think we need these things. Explain that mining, or extracting resources, provides us with valuable materials or products that we use everyday, such as fuel for our cars, energy to heat our homes, raw materials for manufacturing things, or precious metals for coins or jewelry.

2. Explain that mining engineers design the methods and tools used for locating and extracting these resources. These engineers are also concerned with designing ways to prevent and/or repair the damage that is done to the environment as a result of mining activities.

3. Hold up two cookies, one soft and one hard, and explain that these chocolate chip cookies represent two different land types, each with a limited supply of a valuable resource—chocolate chips. Tell families that they are going to be a team of engineers and miners, hired to design and implement a process for mining the chocolate chips out of a cookie. The company has asked that you do the least damage to the cookie as possible during mining, because it will cost them to repair the cookie after the mining is done.

4. Distribute the *Mining for Chocolate Design Challenge* and a bag of supplies to each family, asking that they leave supplies in the bag for now. Review the design challenge, pointing out that engineers don't just design things, they also design methods and processes for getting a job done. Ask families if they have any questions about the challenge.

5. Have families spend about 5 minutes discussing and making plans for how they will mine the chocolate chips. Tell families to consider which tools might be best for the two different landmasses—soft and hard cookies—and how they might do the least damage to each cookie when extracting the chocolate chips.

6. Before handing out cookies, remind families that they should not eat the cookies. Have families place the paper towels from their bags on the table. Distribute one soft cookie and one hard cookie to each family on the paper towels. Ask families to NOT begin their mining yet, but rather to examine the two cookies and estimate the number of whole chocolate chips they expect to extract from each cookie within a 10 minute timeframe. Explain that mining engineers and geological engineers would analyze and test the landmass to predict how much of a resource is present and decide whether the landmass is a good place to mine. Have families record their estimates on the data sheets. Remind them to keep the chocolate chips extracted from each cookie separate using the two labeled paper plates, so that they can see how close they came to their original estimates. If plates are not labeled yet, have families label one "hard" and the other one "soft" using their pencils.

EXPLOSIVE INVENTION

Dynamite is often used in mining. Alfred Nobel is credited with the invention of modern dynamite, for which he received a patent in 1867. Alfred Nobel donated much of his money to create the Nobel Prizes in 1900.

7. Announce to the group that it is time to try out their mining tools and techniques. Remind families that engineers are always looking for ways to improve their designs, so they should feel free to change the mining process they have chosen as they go, or as they look around and see what is working well for others. Signal for families to begin mining—announcing that they will have exactly 10 minutes to extract as many chocolate chips as possible.

8. When the 10 minutes is up, call for everyone to stop mining and put down their tools. Have families count the number of whole and partial chocolate chips extracted from each type of cookie, record these amounts on the data sheet, and compare the count with the estimates made prior to the extraction. Discuss the reasons for any discrepancies between the two numbers. Ask the group if one land-type (cookie material) was easier to mine than the other. Did one land type yield more whole chocolate chips? Why? Explain that a mining engineer needs to choose the best place to mine. Engineers need to think about how much of a resource is in a particular place, and how much it will cost to either mine the resource from the land so that the land is not damaged or to restore the land after mining.

9. Explain that it is time to count up the earnings and associated costs from their work. Using the chart on the back of the *Design Challenge*, have each family add up the total value of their extracted resources (whole and broken chocolate chip pieces) and record this on their data sheet.

ENGINEERING CONNECTION

Engineering is involved in many aspects of mining. A geological engineer finds the mineral. A mining engineer figures out how to get the resource out of the ground safely and with minimal harm to the environment. A chemical engineer develops the process to separate the valuable resource from the rock so that it can be used. A civil engineer designs the roads and buildings at the mine. A safety engineer finds ways to protect the miners' health and safety. An environmental engineer tests the mine's air and water quality, as well as designing ways to reclaim the land—put it back to as close to its original condition as possible.

10. Remind the group that there are costs associated with environmental repair which will reduce their profits from the sale of the extracted resource. Have families set aside whole chocolate chips or chocolate chip pieces from their pile based on the "Cost of Environmental Repair" chart on the back of the *Design Challenge*. Have them add up what these chips are worth, and record this on their data sheets. Then have families subtract the cost of environmental repair from the value of the chocolate chips extracted. This is their net profit from the mining process.

11. Ask families to share which extraction techniques were most effective. Which were most destructive? Was their cookie in pieces after the 10-minute extraction? What changes would they make to their mining process if they tried it again? How do they think engineers solve the problem of "repairing" the earth's surface after mining?

Extensions

▶ Suggest that families try mining for items in other foods at home, such as seeds out of a watermelon, or raisins from a granola bar. Design tools and mining processes for each situation that can extract the materials with the least amount of damage to the "land."

MAKING REPAIRS

Restoring the land and environment after impact from a mining operation is called **reclamation**. This may include returning rock and soil to an open pit mine, removing waste or pollution generated by the mining activities, and covering the area with healthy topsoil and vegetation.

A mountaintop coal mine in Nicholas County, West Virginia before reclamation.

Another mountaintop coal mine in Logan County, West Virginia after reclamation.

STOP AND THINK

Engineering Fields
- *materials engineering*
- *general engineering*

Engineering Concepts & Skills
- *properties of materials*
- *engineering design process*
- *invention/innovation*

Estimated time: 20-30 minutes

General Supplies
- *mug with a handle*

Supplies Per Family

(one set for **each** family or small group)

- *Stop and Think Questions (Appendix D)*
- *lunch-size brown paper bag*
- *various everyday objects (1 different object for each bag)—pen, child scissors, spoon, key, straw, felt-tip marker, water bottle, bowl, comb, toothbrush, fingernail clippers, small zip-lock bag, fork, whisk, wooden spoon, spool of thread, small trowel, paintbrush, pet toy, bolt, funnel, small squeegee, bungee cord, screwdriver, pencil, etc.*

We use engineered objects everyday without giving much thought to how they were designed. In this activity, family members will "stop and think" about some familiar everyday items. Focusing on an object's purpose, design, and materials, families will share their ideas with each other and then shift their thinking to brainstorm new and creative uses for the object.

Advance Preparation

- Gather objects and place one in each bag. Estimate the number of families involved and prepare that many bags plus several extra. Fold the top of each bag to hide the object inside.

- Copy the *Stop and Think Questions* (one per family).

HUMAN FACTORS ENGINEERING

When designing a product, engineers often must think about how to make that product "user-friendly." User-friendly design decisions might include making the handle of a mug fit a person's hand or making sure a container of hot liquid does not burn a person's hand. When engineers need to consider human needs, or how people will interact with a product or process, this is called "human factors engineering."

Activity Steps

1. Hold up the mug and, by asking the families, come to consensus that it is in fact a mug, a common item found in most homes.

2. Suggest that everyone now "stop and think" in a different way, examining the mug in a way that engineers might. Involve the group in answering the following questions:

   - For what purpose or use did engineers design this object? *(hold hot beverages)*

   - What design features were used to make this object work? Think about the material it is made of and the shape of the object. *(It is made with material that is rigid and waterproof; won't leak . . . a floppy material would spill the drink, etc.) (The handle makes it easier to hold if the drink is hot; if it were shaped like a bowl it would spill; etc.)*

   - What other uses could this object have? Think creatively. Practical and fantastical ideas are welcome! *(Use as a pencil holder or paperweight . . . place over your ear to muffle sound . . . flip over and use bottom of mug as a drum . . . etc.)*

3. Explain that families will be playing a game called Stop and Think. Hold up one of the paper bags, explaining that each family will receive a paper bag with an object inside. Families will discuss the object found inside the bag, using the same questions that the group just answered about the mug. The questions are listed on the *Stop and Think Questions* and family members should take turns asking the questions.

4. Establish a signal, such as a hand clap, whistle, or turning lights off and on, and tell the group that when they get the signal they are to put the object back into the bag, fold the top of the bag over, and pass the bag to another family.

5. Distribute the *Stop and Think Questions* and one closed bag to each family. Set aside the extra bags nearby. Give families 3-5 minutes to discuss the object found in their bags.

6. After 3-5 minutes, use the signal to stop discussion. When all the objects have been returned to their bags, have families stay in their seats and pass the bags around until the words, "Stop and Think" are announced, at which time they should open the bag that they are currently holding and use the three questions to discuss their new object.

ENGINEERING CONNECTION

Engineers design objects to meet human needs or solve problems. **Convergent thinking** is used to narrow ideas down and "converge" on a single design solution. **Divergent thinking** is used when a variety of ideas are brainstormed, starting with a single object or design solution and trying to "diverge" from the original design and consider alternative uses for the same design.

7. Continue in this manner for 3 or 4 rotations of 3-5 minutes each.

8.   Use a final signal to stop family discussion. This time, announce that families should keep their objects out of the bags. Ask families to team up with one or two other families near them. Explain that their last challenge will be to think of ways to combine all their objects to create a new product with a new purpose. After a few minutes, allow time for families to share their new combined products with the whole group.

A STICKY SOLUTION

The Post-It® Note is an invention that began as a solution to a simple problem. Art Fry, a chemical engineer at 3M, sang in a choir. He wanted to find a way to mark the pages in his songbook with a bookmark that didn't fall out on the floor. It had to stick to the paper and yet be movable without tearing the paper. Another scientist at 3M had created a new adhesive material in the form of small sticky spheres. Art Fry was able to bond these spheres to paper with a specific spacing between the spheres, creating a sticky bookmark that could be removed without damaging the pages of a book. He also used some of his sticky bookmarks to write notes to his boss. This broadened his original idea for the product and the Post-It® Note was born!

Extensions

▶ Invite families to take an extra *Stop and Think Questions* sheet with them to play Stop and Think at home.

▶ Families can create a commercial or advertisement for a new "product" that focuses on the "engineered" features of that product.

▶ Play Stop and Think using not-so-familiar objects from other countries, such as a molinillo (wooden froth-making tool for Mexican hot chocolate), or historical objects such as a pocketwatch.

Families "team up" to build a structure using only their collective creativity, effective communication skills, and pipe cleaners!

Engineering Fields

- *general engineering*

Engineering Concepts & Skills

- *teamwork*
- *communication*
- *engineering under constraints*

Estimated time: 20-30 minutes

Supplies Per Family

(one set for **each** family or small group)

- *20 pipe cleaners*
- *quart-size resealable plastic bag*

Advance Preparation

- Place 20 pipe cleaners in each resealable plastic bag for easy distribution.

- *Event tip:* if you would like to reuse the pipe cleaners, ask families to disassemble the structures at the end of the activity and straighten out their pipe cleaners before returning them to the resealable bag.

Activity Steps

1. Invite families to imagine that they are teams of engineers who have been hired to design the framework for the model of a tall building. The model will be built out of pipe cleaners. Family members will work together as a team to create the tallest free-standing (i.e. not hand-held or attached to anything) tower with only the materials provided.

2. Tell families that engineers are often faced with constraints that can make a project more challenging to complete. Explain, for this challenge, that one of their constraints will be time. They will have just 12 minutes to complete the challenge. Explain to the group that during the challenge you will give a signal (i.e. whistle or clap), which means that everyone needs to stop working and listen for new information before continuing with the challenge.

3. While you pass out the bags of pipe cleaners, ask families to leave the pipe cleaners in the bag as they spend a few minutes planning how they will construct their towers. Planning is an important part of the engineering process. After allowing for planning time, give the signal for families to take out the pipe cleaners and begin assembling their towers.

4. After about two minutes, give the signal to stop and announce the following: *"We have just received a notice from the finance department. Our budget has been cut. Effective immediately, we now have only 15 pipe cleaners with which to build our towers. Please put five pipe cleaners back in the bag and return to work."*

5. After three more minutes, give the signal again and announce the following: *"We have just found out that a portion of this project is now being managed by a company from another country. Their engineers do not speak the same language that we do. So, starting now, ALL team members are forbidden to speak or write and your team must find other ways to communicate. You may return to your work."*

6. After three more minutes, repeat the signal and announce the following: *"We have just received a memo from the human resources department at our engineering firm. We have to reduce our workforce by 50%! Starting now, EACH team member must work with one hand behind his or her back. All the other reductions are still in place, so still limit your team to 15 pipe cleaners and no talking! You may return to work."*

7. After four more minutes have passed, give the signal and announce that time is up. Congratulate all teams on the success of building some sort of tower (regardless of height) and recognize a few of the differing designs around the room.

ENGINEERING CONNECTION

Many people think engineers spend much of their time working alone at a desk or in a lab. The fact is that engineers are often trained to work in teams. Each individual engineer supplies vital information and expertise that contributes to the success of the entire project. Successful teams communicate effectively, making it possible to respond to changes in specifications or requirements by collectively working through the constraints.

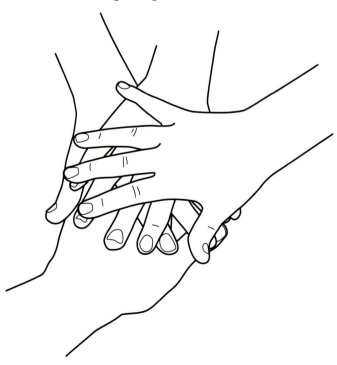

8. Ask a few families to share their experiences dealing with the different constraints that were introduced during the activity. Have them describe how they were affected by the constraints. For example, what was the most difficult constraint for them? How did they overcome it? What would they do differently if this activity were repeated?

9. Explain that engineers often work on a team to solve a problem or meet a challenge. Effective communication skills and the ability to work well with others to overcome unexpected challenges and constraints are important skills for being a successful engineer.

"HOUSTON, WE'VE GOT A PROBLEM"

In 1970, this calm announcement was made by Apollo 13 astronaut Jack Swigert to NASA's Mission Control. Swigert then informed NASA of an on-board explosion, creating the potential for the astronauts to run out of air. This started a frenzied few days of engineering problem-solving that was instrumental in bringing the three astronauts home safely. In less than two days, the NASA engineers devised a way to retrofit on-board chemical containers to clean the air and provide oxygen for the men to breathe on the way home. The solution had to use only the limited materials available to the astronauts in space, such as plastic bags, cardboard, and duct tape, and then it had to be communicated to the astronauts for in-space assembly. Teamwork, innovation, and communication were vital components of making this engineering 'miracle' succeed.

APPENDICES

The materials provided in the appendices may be reproduced for the purposes of planning and implementing Family Engineering events in educational and community settings. When copying these items, we recommend that the spiral edge appearing on copies be trimmed off to create a cleaner presentation. For use with multiple events, sets of durable, laminated materials, including sign holders, are available at **www.familyengineering.org**.

OPENER ACTIVITY SIGNS

AGAINST THE WIND

How can engineers save energy through design?

1. Turn on the fan and hold one car at the top of the ramp. Let go of the car, allowing it to roll straight toward the fan. What happens?

2. Try again with another car until you have tested all the cars. What do you notice?

3. Which car is the most aerodynamic (moves easily through the wind)? Which car do you think would need to use the most energy (fuel) to move against the wind?

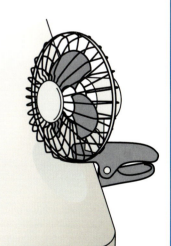

Want to know more? See back of sign.

ENGINEERING CONNECTION

At highway speeds, most of the energy (fuel) needed to keep a car moving down the road is used to push air out of the way. Engineers can help us save energy by designing more **aerodynamic** cars and trucks. This means that they have minimal air resistance and move through the air easily.

Some strategies for making a car more aerodynamic include changing the shape of the car, making rearview mirrors smaller or placing them inside the car, covering the wheel openings, and lowering the car so that it is closer to the ground.

This car is designed to be aerodynamic and will experience less air resistance as it moves through the air.

This car **is not** designed to be aerodynamic and will experience a lot of air resistance as it moves through the air.

ALL THE RIGHT TOOLS

Can you choose the best tool for a job?

Engineers rely on accurate measurements to do their work. Which measuring tool will work best to accurately measure each of the items below?

- Width of the table

- Length of your finger

- Small amount of rice

- How long it takes to walk around the table

- Width of your thumbnail

- Large amount of rice

- Length of the room

- Length of your arm from elbow to wrist

- Your height

Want to know more? See back of sign.

ENGINEERING CONNECTION

Taking accurate measurements is very important in engineering. Engineers use accurate measurements to help them draw a design, construct models, or give instructions to the people who will create or build a product the engineer has designed.

ARCHES

Are arches just "artsy" or do they actually make a bridge stronger?

1. Place the cans in the two circles on the paper. Then place the paper strip across the cans to form a **flat** bridge.

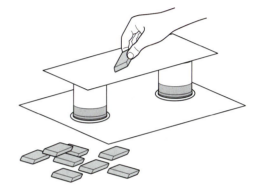

2. How many erasers can you place on the center of the flat bridge before it collapses?

3. Place the ends of the paper strip between the two cans to form an **arch**. How many erasers do you think this bridge will hold? Try it out!

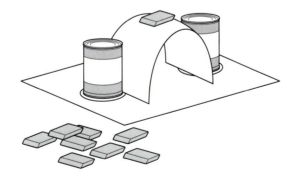

Want to know more? See back of sign.

ENGINEERING CONNECTION

Arch bridges get their strength from their shape. Instead of pushing straight down, the weight of the bridge, and any objects on the bridge (the load), are carried outward along the curve of the arch to the supports at each end.

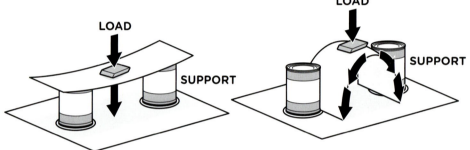

Engineers may choose an arch shape in order to have a larger space between supports, providing more room for water to flow or boats to pass beneath the bridge.

Concrete arch bridge near Hoover Dam.

Pont du Gard aqueduct (a structure to provide water) in France was built above an arched bridge over 2,000 years ago.

BOXING BEANS

Are packages engineered?

1. Predict which package shape will hold the most beans.

2. Fill one package to the top with beans so that the lid can still close.

3. Pour the beans directly into another package. Does this new package hold more or less?

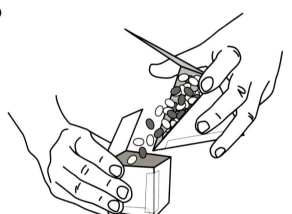

4. Repeat with all of the packages. What did you discover?

5. If you were an engineer, which package shape would you use to:

   - Make it appear that it holds the most product?
   - Stack easily on a shelf?
   - Attract attention with a unique shape?

Want to know more? See back of sign.

© 2011 Family Engineering

ENGINEERING CONNECTION

Surprise! All the package shapes hold about the same number of beans. Why do engineers decide to use one shape instead of another?

Engineers put a lot of thought into developing packaging for the products we buy—which materials to use, how much the package costs to make, is it attractive, does it protect the product, what is the impact on the environment,  and much more. With more and more products being created, package engineering is a field with lots of opportunities!

DOMINO DIVING BOARD

How can engineers help us "hang out" safely?

1. Build a ledge that "hangs out" over the edge of the book, like a diving board over a pool. **No dominoes can touch the table!**

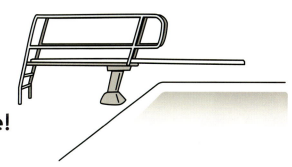

2. Watch the ruler to see how far your ledge "hangs out" before it collapses.

3. Improve your design and try again!

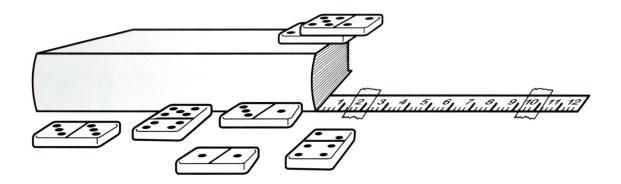

Want to know more? See back of sign.

ENGINEERING CONNECTION

A **cantilever** is a structure that is connected to a support at one end and extends out beyond support on the other end. Engineers must design cantilevers to be structurally safe. The fixed end must have enough support, or weight, to hold up the weight of the other extended end. Some examples of cantilevers are diving boards, balconies, and airplane wings.

COLOSSAL CANTILEVER!

The Grand Canyon Skywalk is a glass-bottom, horseshoe-shaped cantilever platform projecting 70 feet out over the cliff edge. It allows visitors to look straight down, nearly 4,000 feet below, to the bottom of the Grand Canyon.

GLUE IS THE CLUE

Can an engineer make something stronger?

1. Stack the two **unglued** cards together and place them across the top of the two cups like a bridge. How many washers do you think the cards will hold before collapsing?

2. Carefully stack the washers in the circle on the top card, counting them as they are added. Keep going until the cards collapse.

3. Now repeat with the **glued** cards. How many washers do you think the **glued** cards will hold?

4. Which pair of cards held more washers? Why?

Want to know more? See back of sign.

ENGINEERING CONNECTION

The two **glued** cards stay together, becoming a single, thicker card with greater strength. The two **unglued** cards do not work as well together to hold up weight because they can bend and slide apart.

Gluing layers of material together is called **lamination**. Plywood is an engineered wood product that uses a special glue developed by chemical engineers to hold thin layers of wood together. This makes plywood stronger than solid wood of the same thickness.

Skateboards are made out of plywood so that they can be both strong and lightweight.

HAPPY FEET

Would you play soccer in rain boots?

1. Match an Activity card with a shoe and place them both on the mat.

2. Select three Design Requirements that an engineer should consider when designing a shoe for this activity. Place these cards on the mat.

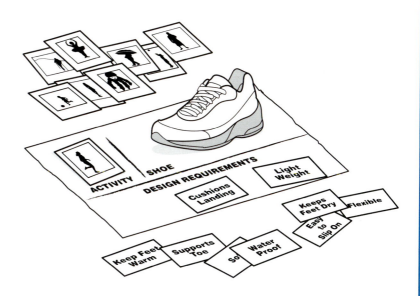

3. Repeat with a different Activity card and shoe.

4. Look at the shoes on your feet! What activity were they designed for?

Want to know more? See back of sign.

ENGINEERING CONNECTION

Designing shoes for specific activities and conditions requires several kinds of engineers. Biomedical engineers help create a design that  is good for the feet and can improve performance; materials engineers design and combine materials to meet special requirements (waterproof, soft, flexible, etc.); and industrial engineers design factory equipment to make the shoes.

INSPIRED BY NATURE

How have sharks and termites helped engineers?

1. Select a Human Invention card.

2. Next, select a Nature's Inspiration card to match this invention.

3. After all the cards have been matched and placed on the mat, read the backs of the Nature's Inspiration cards to find out how engineers are inspired by nature.

4. Can you think of other engineered inventions that might have been inspired by nature?

5. Remove the cards from the mat when you are finished.

Want to know more? See back of sign.

ENGINEERING CONNECTION

Sometimes engineers come up with a new idea for an invention by observing something in nature.

All of the inventions on the Human Invention cards have been inspired by something in nature. The story of the inventions can be found on the back of the Nature's Inspiration cards.

Next time you're outdoors, take some time to look around and wonder.

LEARNING FROM FAILURE

How can failure lead to success?

1. Create a boat out of **one** piece of aluminum foil and place it in the water. Predict how many pennies you think your boat will hold before it fails and sinks.

2. Place pennies in your boat gently, one-by-one. Watch the boat carefully as it gets close to sinking.

3. Can you change your boat design to hold more pennies? Try again using the same foil or **one** new piece.

4. What did you learn from watching your boat sink?

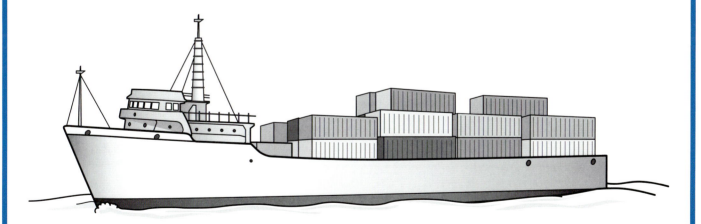

Want to know more? See back of sign.

ENGINEERING CONNECTION

Failure plays an important role in the design process. Engineers use failure to help them find better solutions. Often, engineers will test a design until it fails in order to see where improvements are needed.

By testing your boat until it sinks, and watching closely, you see where your design can be improved to keep the boat floating longer or to help it carry a heavier load of pennies.

FAMILY engineering

LET'S COMMUNICATE

As an engineer, how well can you explain your design to a builder?

1. Face a partner so that you each have a box on its side and an identical set of blocks in front of you.

2. One of you will be the **engineer.** Design and build a structure inside of your box. Do not allow your partner (the builder) to see your structure.

3. Next, have the **builder** try to build a copy of the engineer's structure in their own box by following the engineer's directions only. The builder cannot ask questions and you cannot look inside each other's boxes.

4. When finished, compare the two structures. How well did the engineer communicate to the builder?

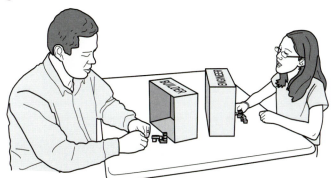

Want to know more? See back of sign.

© 2011 Family Engineering

ENGINEERING CONNECTION

Communication is harder than we think! To be successful, an engineer must be able to communicate effectively with others and work well on a team. Accurate and clear communication between all team members is key to a team's success.

MAKE IT LOUD!

How can an engineer turn up the sound?

1. Hold the open end of one tube to your ear.

2. Have **someone else** tap lightly on the covered end. What do you notice?

3. Repeat using the other tubes, keeping the same level of tapping.

4. Which material makes the tapping sound louder? Which material makes the tapping sound quieter?

Want to know more? See back of sign.

ENGINEERING CONNECTION

Biomedical engineers design devices and procedures that solve medical and health related problems. These devices include artificial organs, artificial limbs, advanced imaging machines, and many other medical instruments and tools that help doctors diagnose and treat their patients.

One tool that is designed to make sounds louder is a **stethoscope**. Stethoscopes are used by doctors to make the sounds inside of people and other animals easier to hear. Listening to these sounds can help a doctor or veterinarian diagnose a possible medical problem. Stethoscopes are also used by mechanics to listen to the sounds made by machines or car engines.

PICTURE THIS

Can you "see" in 3D?

1. From the four examples, choose a 3D shape you would like to make.

2. Select a "Design Page" you think will make this shape.

3. Cut out your shape by cutting along the dotted lines only.

4. Fold along the solid lines, then use tape to complete your 3D shape.

5. Does your 3D shape look the way you thought it would?

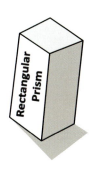
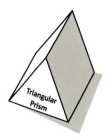
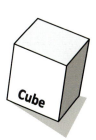

Want to know more? See back of sign.

ENGINEERING CONNECTION

When engineers design something, they usually draw a two-dimensional (2D) version on paper or with a computer before assembling a three-dimensional (3D) model. The ability to imagine or "see" how a 2D picture can become a 3D object is called **spatial visualization**, a very useful skill in engineering!

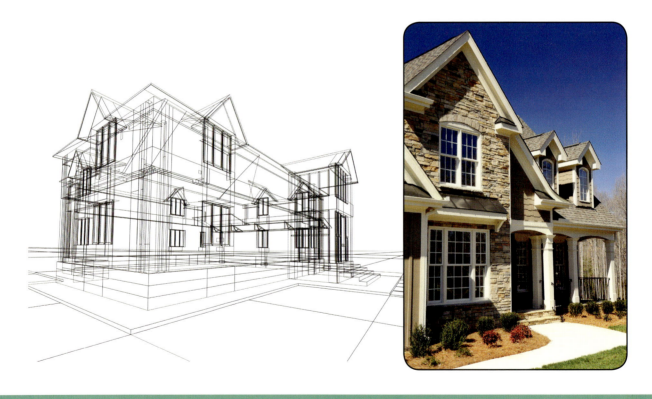

SHIFTING SHAPES

What shape does an engineer favor for stability?

1. Select a square and a triangle. Gently try to move the sides of each shape up and down.

    - Which one is more rigid and keeps its shape?

    - Which one shifts and changes shape easily? Can you add another strip to this shape to make it more stable?

2. Try making a 5-sided shape. Does it shift and change its shape easily? What can you do to make it more stable?

3. Before you leave, please take apart any extra shapes or added fasteners and strips, leaving only the original squares and triangles at the table.

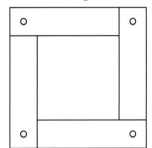 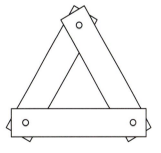

Want to know more? See back of sign.

ENGINEERING CONNECTION

When engineers design a structure, it needs to be stable and keep its shape. Engineers often add strength and stability to the structures they design by using materials in the shape of a triangle. Triangles don't twist, bend, or collapse easily. If you look closely, chances are you will see lots of triangle shapes in the structures around your community.

SHOWERHEAD SHOWDOWN

When is less actually more?

1.   Look at the bottom of the two cups. How are they different? Which one do you think will drain water the fastest?

2.   Do the 5-second shower test. Start by holding the cups upright by their open rims.

3.   Lower both cups straight down all the way under the water so they fill to the very top (see below).

4.   Lift both cups out of the water **at the same time**, holding them over the tub to drain. Count to 5 slowly.

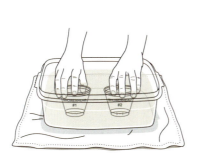

5.   Which cup (showerhead) uses the **least** amount of water for a 5-second shower? Which showerhead design would help you save water at home? Why?

Want to know more? See back of sign.

ENGINEERING CONNECTION

In the United States, showering consumes about 20% of the total water used by a family in an average home. Older showerheads have fewer but larger holes for water flow. Newer "low flow" showerheads are designed with a greater number of smaller holes. "Low flow" showerheads are engineered to use less water, save energy, and still provide a nice shower.

SOLID GROUND

How firm is your foundation?

1. Place the block **on top** of one earth material.

2. Place two fingers in the middle of the block and press down. What happens?

3. Repeat with the other two earth materials. Does the block push easily into each material?

4. How are the materials different? What do they feel like?

5. If you were an engineer, which material would you use to provide a firm foundation for a house? Which material would you use under a playground, to cushion a fall?

Want to know more? See back of sign.

ENGINEERING CONNECTION

Geological and civil engineers test different types of earth materials to determine a good place to build. Materials such as crushed rock, where the pieces are rough and do not slide against each other easily, can lock together and support more weight or pressure. This is a good material to use as a foundation under a building or a road.

A house should be built on a firm foundation.

A play ground also needs a firm foundation, but the material covering the surface needs to cushion a fall.

Smooth, rounded particles, like sand or rounded pea gravel, can easily slide against each other and move easily around other objects. This creates a surface that is not a good foundation for heavy construction, but can be good for absorbing impact and cushioning a fall.

SOUNDPROOF PACKAGE

Can you engineer a package that reduces noise?

1. Shake a noisemaker to hear the noise it makes.

2. Design a package for the noisemaker that will muffle the noise. Use one or more of the materials provided and package the noisemaker in the plastic container. Attach the lid.

3. Test your package by shaking it. Can you still hear the noisemaker?

4. Try again with different materials. Which materials make the best soundproof package?

5. Please take apart your package and separate the materials when you have completed your tests.

Want to know more? See back of sign.

ENGINEERING CONNECTION

Noise is unwanted sound. Engineers design products and materials to protect the human ear from loud noises and reduce or remove unwanted sounds.

People wear ear protection when working around loud noises.

THRILL SEEKERS

Can you engineer a good roller coaster design?

1. Work as a team to hold the tubing in a roller coaster shape.

2. Put a marble into the top of the tube and watch it ride. Did it make it to the end?

3. How many loops and turns can you add to make the most thrilling ride and still get the marble to the end?

Want to know more? See back of sign.

© 2011 Family Engineering

☀ ENGINEERING CONNECTION

Mechanical engineers designing a roller coaster must balance the thrill of the ride with safety. They use the science of how things move to design hills, loops, twists, and turns that give riders a safe but thrilling ride!

The idea of 'coasting' for fun comes from ice slides popular in Russia in the 17th century. Huge wooden structures with a thick sheet of ice allowed people to climb a stairway, get on a sled, and then careen down a steep ramp and up an opposite ramp, going back and forth before eventually stopping in the middle.

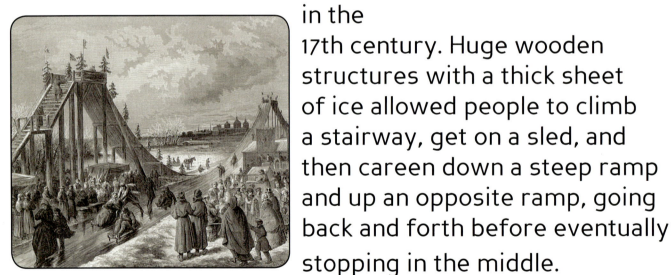

TUMBLING TOWER

How do engineers keep tall structures from tumbling?

1. Use 14 tubes and 3 cardboard squares to build this tower.

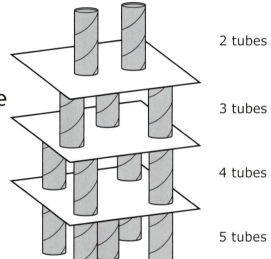

2 tubes

3 tubes

4 tubes

5 tubes

2. Following the rules below, take turns removing one tube at a time from the tower without letting the 3 cardboard platforms fall.

   - You may use both hands.

   - You may touch the cardboard platforms only when removing or moving a tube.

   - You may change the position of the remaining tubes.

3. What steps did you take to keep the tower from tumbling? Why did the tower eventually fall?

Want to know more? See back of sign.

ENGINEERING CONNECTION

Engineers often test a structure's strength until it fails and then try to figure out why the failure happened. The weight that a structure supports is called its **load**. In order for a structure to be stable,

the load must be balanced. If a structure is changed in some way, the load may need to be rebalanced to maintain stability.

WHAT DO ENGINEERS DO?

What do *you* think engineers do?

1.   Read the statement on each card.

2.   Place one bean into each cup that you think describes what an engineer does.

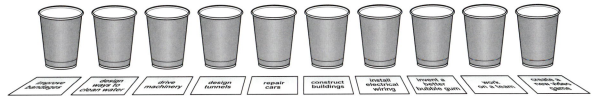

improve bandages | design ways to clean water | drive machinery | design tunnels | repair cars | construct buildings | install electrical wiring | invent a better bubble gum | work on a team | create a new video game

Want to know more? See back of sign.

ENGINEERING CONNECTION

Engineers are creative problem-solvers who play a vital role in society. People often believe that engineers fix or build things and that they only work on buildings, bridges, and cars. Actually, engineers are the people who **design** these and thousands of other products. There are many different engineering fields.

- Some engineers develop new technologies to help sick or injured people feel better, such as new medical machines, medicines, and devices to help people walk, see, or hear.

- Some engineers develop solutions for everyday challenges and design products we use at home, such as can openers, refrigerators, computers, and packaged food.

- Some engineers even help design solutions to the world's biggest problems—like the need for clean drinking water or developing new energy sources.

FAMILY
engineering

WHO ENGINEERED IT?

How many engineers does it take to design a light bulb?

1. Deal the Engineer cards out to family members until all cards are distributed.

2. Choose an Engineered Product card and place it on the game board.

3. Have each family member choose the Engineers from their hand that they think helped design this product and place their cards on the game board.

4. Turn over the Engineered Product card to check your answers.

5. Deal the Engineer cards again and repeat with another Engineered Product card.

6. Please leave the two sets of cards in separate stacks on the table when you are finished.

Want to know more? See back of sign.

ENGINEERING CONNECTION

Products and structures don't just "happen"! They are imagined, designed, and engineered by people trained in an engineering field.

Most products are actually designed by a team of engineers from different engineering fields. For example, to design and produce an automobile requires at least five different kinds of engineers working together: mechanical, electrical, computer, materials, and biomedical engineers.

WRAP IT UP!

Does your house need a raincoat?

1. Which materials will keep a cotton swab dry?

    - Select one square of any material. Wrap it around the end of a cotton swab and hold it in place with your fingers.

    - Dip the covered tip of the cotton swab into the water and count to 10, slowly.

    - Lift the cotton swab out of the water and remove the square of material. Is the tip of the cotton swab dry or wet?

2. Test the other materials with a dry cotton swab. Which material would you use to protect your house from water damage?

Want to know more? See back of sign.

ENGINEERING CONNECTION

When engineers design a building, it is important for them to know how different materials react to water. This knowledge will help them choose water resistant construction materials or add a barrier layer that will keep the water out altogether.

In house designs, engineers usually use black tar paper or plastic sheeting to act as a water barrier to protect the wood underneath.

© 2011 Family Engineering

YOUR FOOT, MY FOOT

Can engineers use their feet for measuring?

1. Have each family member measure the length of the tape line using their own feet. Do this by walking heel to toe along the tape and counting the number of "foot" lengths from end-to-end.

2. Did everyone get the same number?

3. Why is it important to use standard units for measuring?

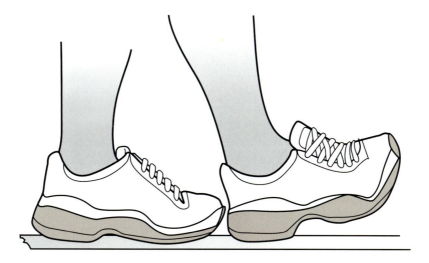

Want to know more? See back of sign.

ENGINEERING CONNECTION

When people talk to each other, speaking the same language helps them to understand each other. This is also true with measurement. It is important that a unit of measurement (such as one foot) means the same thing to everyone. That is why we measure length with a ruler that has a standard length for one foot, instead of letting people measure with their own feet that are all different sizes.

Accurate measurement is essential in engineering. Engineers use accurate measurements to help them draw a design, construct models, or give instructions to the people who will create or build a product the engineer has designed.

OPENER ACTIVITY INTERACTIVES

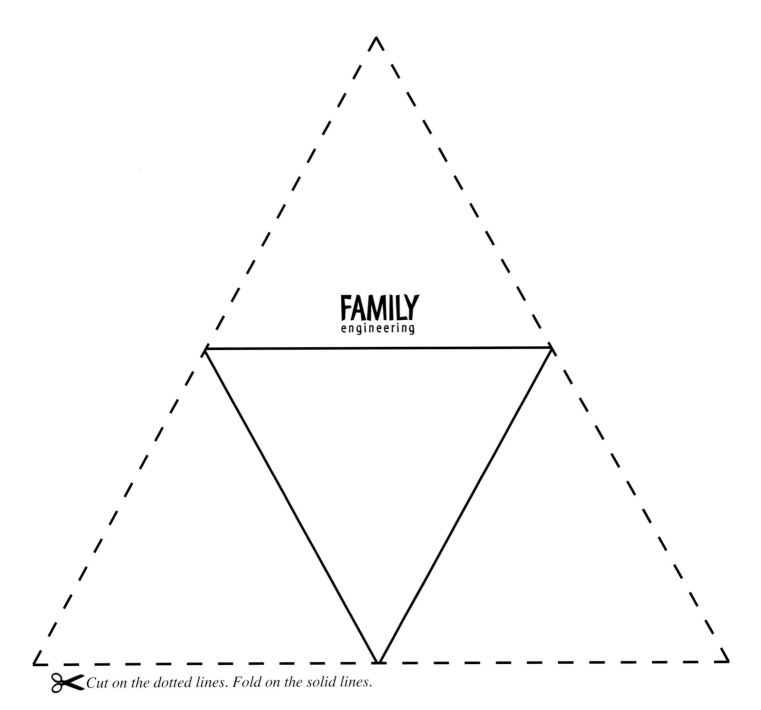

FAMILY
engineering

✂ *Cut on the dotted lines. Fold on the solid lines.*

Boxing Beans: Shape #2

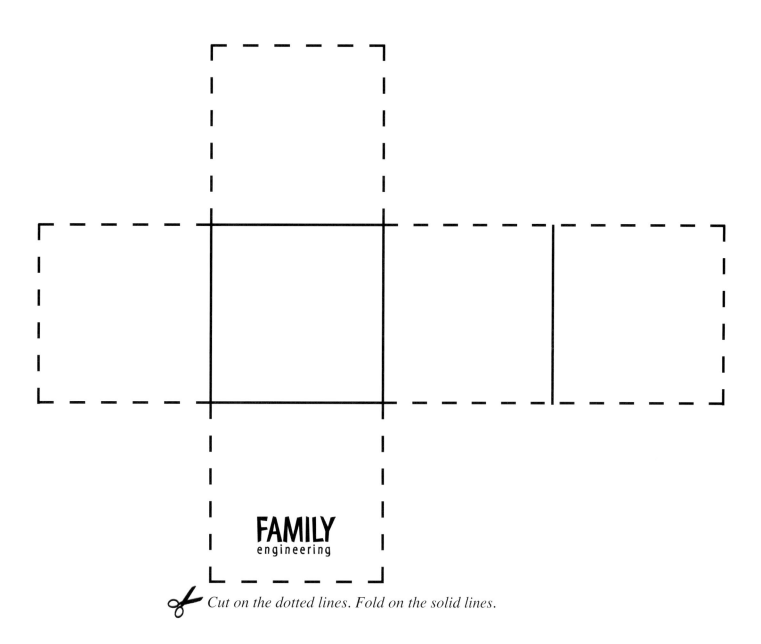

Cut on the dotted lines. Fold on the solid lines.

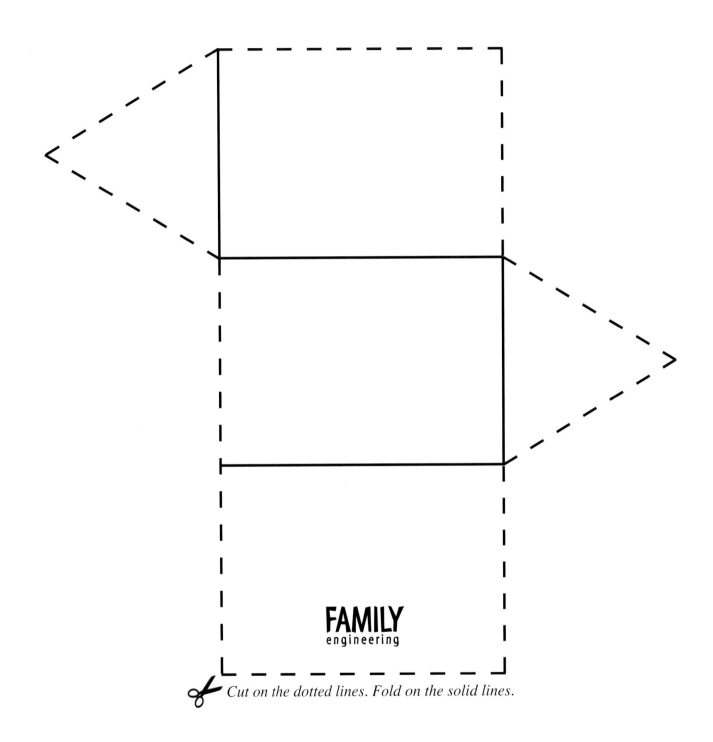

Cut on the dotted lines. Fold on the solid lines.

FAMILY
engineering

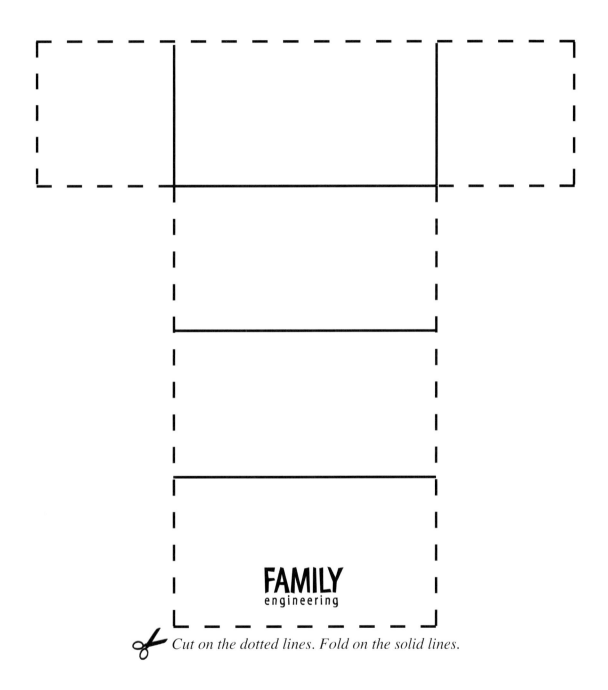

FAMILY
engineering

✂ *Cut on the dotted lines. Fold on the solid lines.*

ACTIVITY

Playing Basketball

ACTIVITY

Running

ACTIVITY

Relaxing at Home

ACTIVITY

Playing Soccer

Copy single-sided. Cut on the guides to create the Activity Cards.

ACTIVITY

Hiking

ACTIVITY

Walking in the Snow

ACTIVITY

Ballet Dancing

ACTIVITY

Walking in the Rain

✂ *Copy single-sided. Cut on the guides to create the Activity Cards.*

ACTIVITY

Walking on the Beach

ACTIVITY

Ice-Skating

ACTIVITY

Roller-Skating

ACTIVITY

Snorkeling

Copy single-sided. Cut on the guides to create the Activity Cards.

Happy Feet: Design Requirement Cards

Soft © 2011 Family Engineering	Supports Ankle © 2011 Family Engineering	Easy to Slip On © 2011 Family Engineering
Lightweight © 2011 Family Engineering	Waterproof © 2011 Family Engineering	Covers Legs © 2011 Family Engineering
Prevents Slipping © 2011 Family Engineering	Cushions Landing © 2011 Family Engineering	Supports Toes © 2011 Family Engineering
Grips the Floor © 2011 Family Engineering	Keeps Feet Cool © 2011 Family Engineering	Lasts a Long Time © 2011 Family Engineering
Keeps Feet Warm © 2011 Family Engineering	Protects Toes © 2011 Family Engineering	Flexible © 2011 Family Engineering

Copy single-sided. Cut on the dotted lines to create the Happy Feet Design Requirement Cards.

HUMAN INVENTION

Velcro

HUMAN INVENTION

Suction Cup

HUMAN INVENTION

Rope

HUMAN INVENTION

Racing Swimsuit

Copy single-sided. Cut on the guides to create the Human Invention Cards.

HUMAN INVENTION

Bullet Train

© 2011 Family Engineering

HUMAN INVENTION

Water-Repellent Fabric

© 2011 Family Engineering

HUMAN INVENTION

Passive Air Conditioning

© 2011 Family Engineering

HUMAN INVENTION

Bionic Car

© 2011 Family Engineering

 Copy single-sided. Cut on the guides to create the Human Invention Cards.

NATURE'S INSPIRATION

Burdock

NATURE'S INSPIRATION

Vines

NATURE'S INSPIRATION

Octopus

NATURE'S INSPIRATION

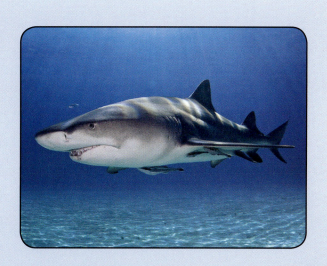

Shark

✂ *Copy back-to-back. Cut on the guides to create the Nature's Inspiration Cards.*

Vines inspired:

Rope

The earliest ropes date back to prehistoric times, and were made from plant fibers, such as vines. The vines were twisted or braided together to form stronger and longer ropes, similar to the way that some vines wrap themselves around a stronger, solid support to continue growing. Engineers have designed many different types of rope for a variety of uses and situations.

Burdock Seeds inspired:

Velcro

In 1948, a Swiss engineer, George de Maestral, took a walk with his dog and came home with plant burrs (seed pods) stuck all over his pants and his dog. After examining these burrs under a microscope, he got the inspiration for creating a new kind of fastener—Velcro! It took eight more years of experimenting to develop and perfect his invention.

Sharks inspired the:

Racing Swimsuit

The V-notch ridges on a shark's skin reduce drag, allowing it to swim fast with less effort. Engineers have designed swimsuits made from materials based on the varying shape and texture of sharkskin. These suits made their debut at the 2000 Olympics in Sydney, Australia and are now commonly used in competitive swimming worldwide.

The Octopus inspired:

Suction Cups

The suction-cup appendages on the legs of an octopus were the inspiration for the modern suction cup, patented in 1882.

NATURE'S INSPIRATION

Kingfisher

NATURE'S INSPIRATION

Lotus Leaf

NATURE'S INSPIRATION

Termite Mound

NATURE'S INSPIRATION

Boxfish

Copy back-to-back. Cut on the guides to create the Nature's Inspiration Cards.

The Lotus Leaf inspired:

Water-Repellent Fabric

A lotus leaf naturally repels water. Engineers have developed a way to chemically treat the surface of fabrics so that they repel water much like the surface of a lotus leaf, making the fabrics more waterproof.

The Kingfisher inspired the:

Bullet Train

A kingfisher bird can dive into water without making a splash. Engineers designed the front of the bullet train to look like the beak of a kingfisher bird so that the train could move through the air more efficiently. When a high-speed train goes through a tunnel, it builds up a cushion of air in front of it that suddenly expands when exiting the tunnel, causing a loud sonic boom. The shape of the bullet train allows it to move through the air in a tunnel without building up that large cushion of air, making it quieter when exiting the tunnel.

The Boxfish inspired the:

Bionic Car

This concept car was designed by Engineers at Mercedes-Benz® to mimic the streamlined profile and sturdy, boxy frame of the boxfish. The bionic car turned out to be stable, fuel efficient, and durable. The company plans to use more of these design elements in future cars.

Termite Mounds inspired:

Passive Air Conditioning

African termites keep their mounds cool by constantly opening and closing vents throughout the mound to direct the flow of air from the bottom to the top. Engineers designed the cooling system of the Eastgate Center in Zimbabwe to mimic the way tower-building termites construct their mounds.

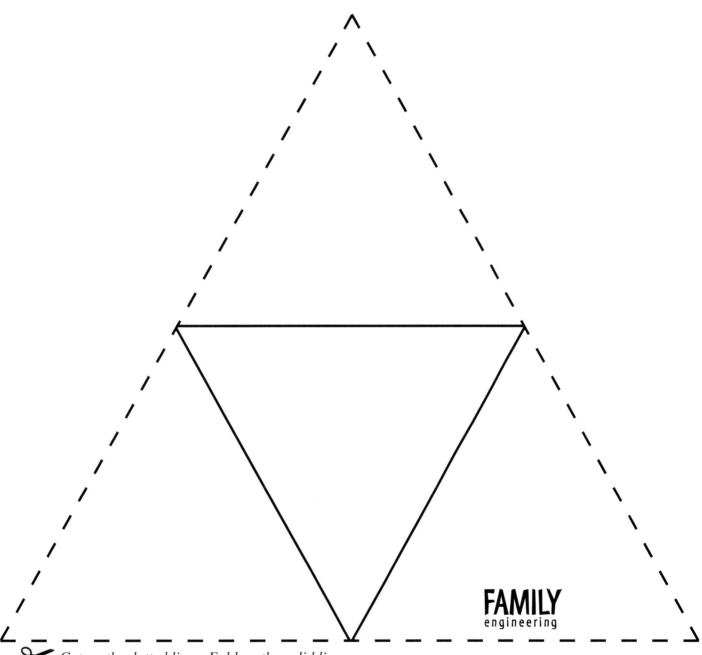

✂ *Cut on the dotted lines. Fold on the solid lines.*

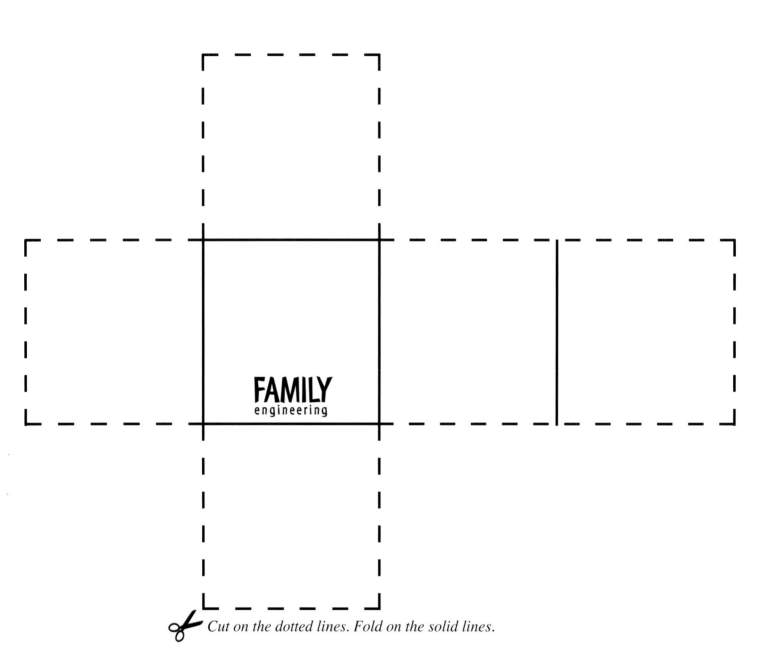

Cut on the dotted lines. Fold on the solid lines.

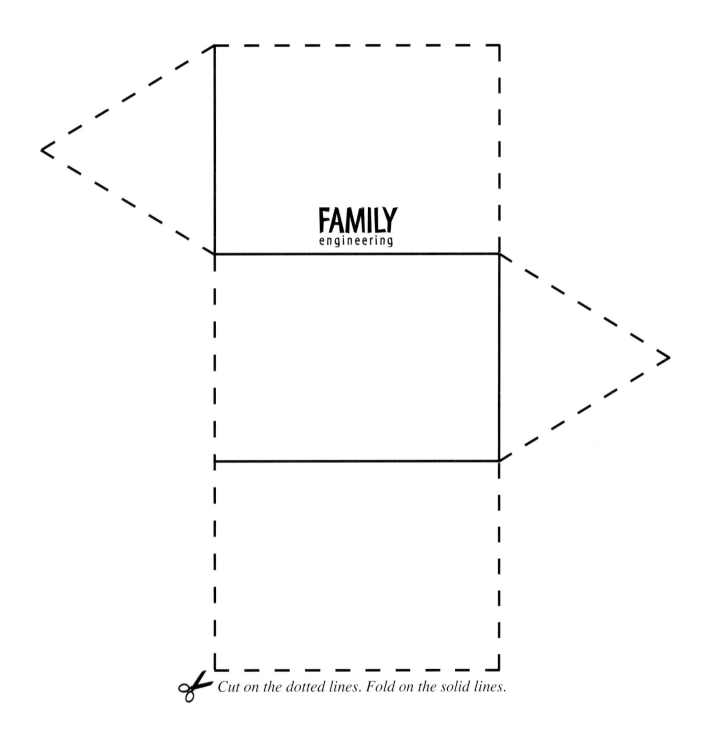

FAMILY
engineering

✂ *Cut on the dotted lines. Fold on the solid lines.*

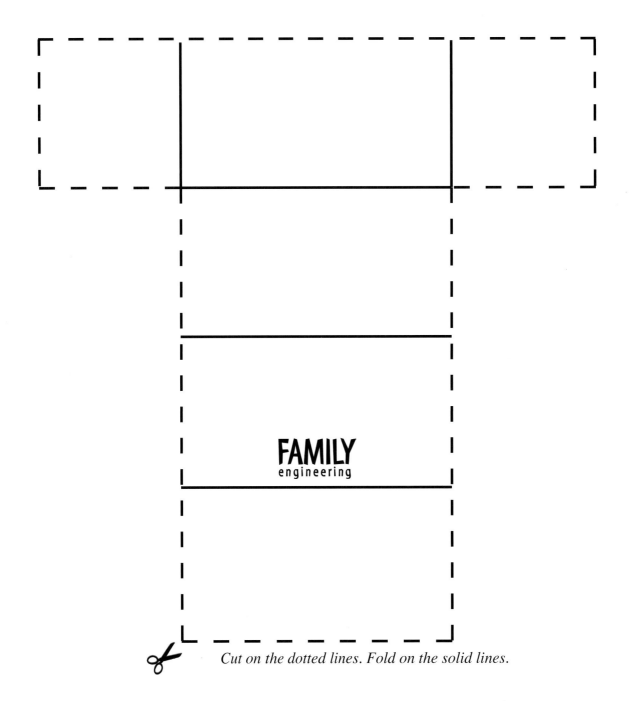

FAMILY
engineering

Cut on the dotted lines. Fold on the solid lines.

Improve bandages

© 2011 Family Engineering

Construct buildings

© 2011 Family Engineering

Design ways to clean water

© 2011 Family Engineering

Install electrical wiring

© 2011 Family Engineering

Drive machinery

© 2011 Family Engineering

Invent better bubble gum

© 2011 Family Engineering

Design tunnels

© 2011 Family Engineering

Work as a team

© 2011 Family Engineering

Repair cars

© 2011 Family Engineering

Create a new video game

© 2011 Family Engineering

Copy single-sided. Cut on the dotted lines to create the What Do Engineers Do? Activity Statement Cards.

BIOMEDICAL ENGINEER

Designs new medical tools, medicines, and methods to improve an individual's health and healthcare.

© 2011 Family Engineering

CHEMICAL ENGINEER

Uses chemicals and other materials to create useful products such as food, medicine, and cleaning products.

© 2011 Family Engineering

CIVIL ENGINEER

Designs structures, such as bridges, roads, buildings, airports, dams, sewage systems, canals, and tunnels.

© 2011 Family Engineering

COMPUTER ENGINEER

Designs computer parts, systems, networks, and software programs.

© 2011 Family Engineering

ELECTRICAL ENGINEER

Designs products or systems that produce electricity or are powered by electricity.

ENVIRONMENTAL ENGINEER

Works to design solutions to environmental problems or challenges, such as providing clean drinking water, reducing water and air pollution, or developing new sources of energy.

MATERIALS ENGINEER

Develops new materials and products to meet a need or solve a problem.

MECHANICAL ENGINEER

Designs machines of all shapes and sizes, from airplanes to toasters, as well as the moving parts of other products, such as toys, furniture, and wheelchairs.

Copy single-sided. Cut on the guides to create the Engineer Cards.

ENGINEERED PRODUCT

Bicycle

© 2011 Family Engineering

ENGINEERED PRODUCT

Video Game

© 2011 Family Engineering

ENGINEERED PRODUCT

Safe Drinking Water

© 2011 Family Engineering

ENGINEERED PRODUCT

Automobile

© 2011 Family Engineering

Copy back-to-back. Cut on the guides to create the Engineered Product Cards.

Video Game

Computer Engineer: writes the computer program.

Electrical Engineer: designs the controller that sends signals to operate the game.

Mechanical Engineer: designs game components and how they work together.

Bicycle

Mechanical Engineer: designs the moving parts.

Materials Engineer: designs the best materials for a lightweight frame or a soft seat.

Automobile

Mechanical Engineer: designs the moving parts.

Materials Engineer: designs shatterproof glass, lightweight car body, and other materials used in automobiles.

Computer Engineer: designs sensors that alert you when the engine is malfunctioning or service is needed.

Electrical Engineer: designs the circuits that power the heating system, radio, and lights.

Biomedical Engineer: designs seat belts, air bags, and headrests to keep people safe.

Safe Drinking Water

Chemical Engineer: designs the process for producing and mixing chemicals that kill bacteria in water.

Environmental Engineer: designs process for testing water.

Civil Engineer: designs the water treatment plant.

Mechanical Engineer: designs equipment inside the water treatment plant.

ENGINEERED PRODUCT

Cell Phone

ENGINEERED PRODUCT

Sports Arena

ENGINEERED PRODUCT

Food Products

ENGINEERED PRODUCT

Prosthetic Leg

✂ *Copy back-to-back. Cut on the guides to create the Engineered Product Cards.*

Sports Arena

Civil Engineer: designs the overall structure of the arena.

Materials Engineer: selects or designs the surface of the sports field.

Electrical Engineer: designs system of lights, loudspeaker, buzzers, and flashing signs.

Mechanical Engineer: designs doors, elevators, escalators, and folding seats.

Cell Phone

Computer Engineer: designs software to operate a phone.

Electrical Engineer: connects the cell phone to satellites, creates ring tones, and flashing lights.

Materials Engineer: selects materials to make a phone lightweight but durable.

Mechanical Engineer: designs factory equipment that makes the phone.

Prosthetic Leg

Biomedical Engineer: designs how the artificial leg operates and fits on the patient.

Materials Engineer: designs materials that look like skin or that are flexible and strong at the same time.

Mechanical Engineer: designs the moving parts of the prosthetic leg.

Computer Engineer: designs sensors to operate the prosthetic leg.

Food Products

Mechanical Engineer: designs the farm equipment and factory equipment to harvest and process the food.

Chemical Engineer: develops ingredients and systems to make and preserve processed and packaged food.

Environmental Engineer: creates best practices and regulations for farms and factories to reduce pollution and waste so that the environment is protected.

ENGINEERING CHALLENGE DESIGN CHALLENGES

ARTISTIC ROBOTS

Design Challenge

Design an "artistic robot" that will draw designs on its own without anyone touching it.

Follow the engineering design process to design and test an "artistic robot."

Robot Design Requirements

▶ Use only the materials provided

▶ Robot must operate on its own, without human assistance, once it is "turned on"

▶ Robot's markings must be visible on paper

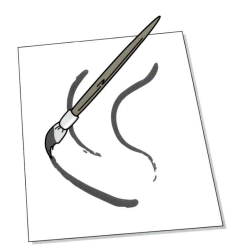

Engineering Design Process

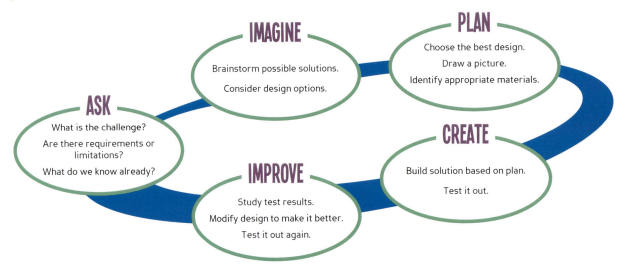

IMAGINE
Brainstorm possible solutions.
Consider design options.

PLAN
Choose the best design.
Draw a picture.
Identify appropriate materials.

ASK
What is the challenge?
Are there requirements or limitations?
What do we know already?

CREATE
Build solution based on plan.
Test it out.

IMPROVE
Study test results.
Modify design to make it better.
Test it out again.

READY FOR MORE?

If time allows, accept a new challenge for your robot. Create a robot design that will make a specific type of design on the paper, such as circles or dotted lines.

FAMILY engineering

BLAST OFF!

Design Challenge

Design a paper rocket that will go the greatest distance when launched with an air launcher.

Follow the engineering design process to design and test a paper rocket.

Rocket Design Requirements

▶ Use only the materials provided
▶ Use an air launcher to launch the rocket

Engineering Design Process

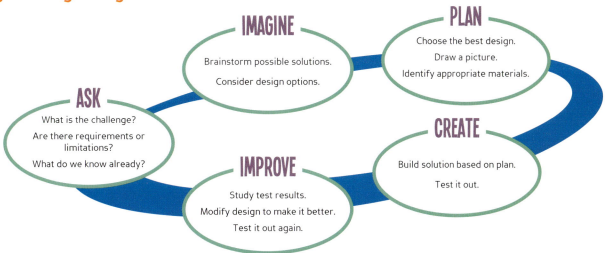

IMAGINE
Brainstorm possible solutions.
Consider design options.

PLAN
Choose the best design.
Draw a picture.
Identify appropriate materials.

ASK
What is the challenge?
Are there requirements or limitations?
What do we know already?

CREATE
Build solution based on plan.
Test it out.

IMPROVE
Study test results.
Modify design to make it better.
Test it out again.

READY FOR MORE?

If time allows, choose a specific distance from the launch pad and try to hit the target. Did you need to change the rocket design in order to hit a specific target?

FAMILY engineering

BRAIN SAVER

Design Challenge

Design a helmet that will protect an egg "head" from injury when dropped from 10 feet off the ground.

Follow the engineering design process to design and test a helmet.

Helmet Design Requirements

- ▶ Helmet must measure 6 inches or less on any side
- ▶ Helmet cannot use a parachute
- ▶ Helmet may cover the entire egg "head"

Engineering Design Process

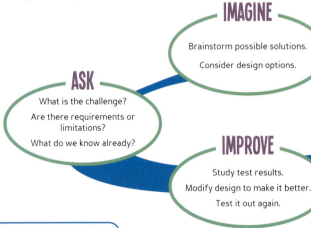

IMAGINE
Brainstorm possible solutions.
Consider design options.

PLAN
Choose the best design.
Draw a picture.
Identify appropriate materials.

ASK
What is the challenge?
Are there requirements or limitations?
What do we know already?

CREATE
Build solution based on plan.
Test it out.

IMPROVE
Study test results.
Modify design to make it better.
Test it out again.

READY FOR MORE?

If time allows, you can:

- Identify the different "engineered" features of your helmet and discuss how each one helps protect your egg "head."

- Add features to your helmet design that would make it attractive to a person buying a new helmet.

FAMILY
engineering

BRIGHT IDEAS

Design Challenge

Sally is on a camping trip and loves to read in her tent before going to sleep. But she can't read and hold the heavy flashlight at the same time. Design a lightweight reading light for Sally that can be attached to her book, her tent, or herself while she is reading.

Follow the engineering design process to design and test a reading light.

Reading Light Design Requirements

▸ Use only the flashlight parts and other supplies provided

▸ Reading light must be hands-free (does not need hands to hold it in place while reading)

Engineering Design Process

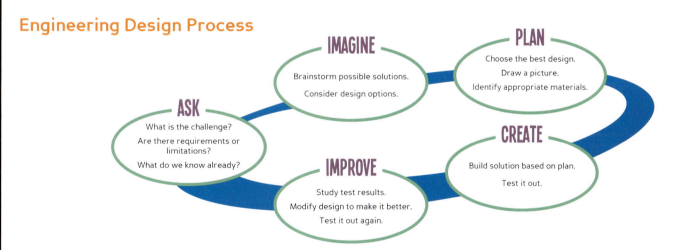

ASK
What is the challenge?
Are there requirements or limitations?
What do we know already?

IMAGINE
Brainstorm possible solutions.
Consider design options.

PLAN
Choose the best design.
Draw a picture.
Identify appropriate materials.

CREATE
Build solution based on plan.
Test it out.

IMPROVE
Study test results.
Modify design to make it better.
Test it out again.

SAFETY NOTES

· Handle the insulated part of the wire only.

· Use masking tape to hold the bare wire in place (not your fingers).

· If a bulb breaks, please inform the activity facilitator so he or she can collect and dispose of the pieces.

FAMILY
engineering

CREATE A CRITTER

Design Challenge

Design an imaginary critter that uses at least two different types of mechanisms for movement.

Moving Mechanism Examples

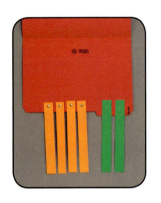

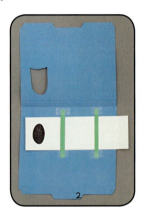

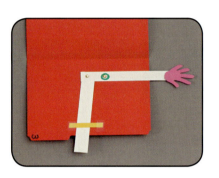

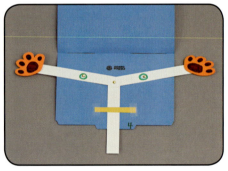

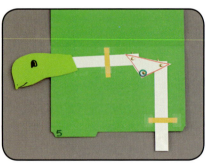

Engineering Design Process

IMAGINE
Brainstorm possible solutions.
Consider design options.

PLAN
Choose the best design.
Draw a picture.
Identify appropriate materials.

ASK
What is the challenge?
Are there requirements or limitations?
What do we know already?

CREATE
Build solution based on plan.
Test it out.

IMPROVE
Study test results.
Modify design to make it better.
Test it out again.

FAMILY engineering

FIVE POINTS TRAFFIC JAM

Design Challenge

> **Design an intersection that allows Tenesha, Juan, and Samantha safe passage as they conduct the following activities.**

1. Tenesha walks to school. After school, she walks to Izzy's Ice Cream and then to the park, before going home.

2. Juan rides his bike to school. After school he rides to the grocery store to pick up some milk for his mom, and then goes home.

3. Samantha is driving on Cherry Street and wants to turn right onto Elm Street to get to the gas station.

Follow the engineering design process to design and test a model of a safe intersection.

Engineering Design Process

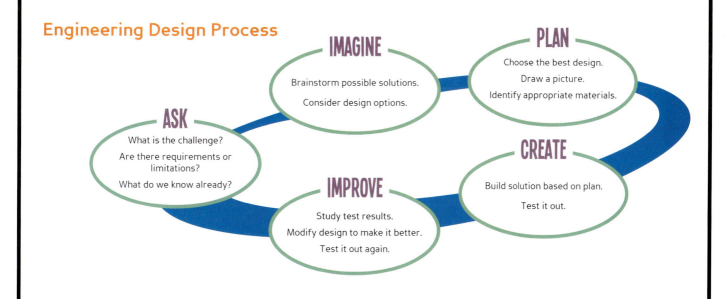

GIVE ME A HAND

Design Challenge

> Design a grabber device that can pick up a cotton ball, an eraser, a pencil, and a marble.

Follow the engineering design process to design and test a grabber device.

Grabber Device Design Requirements

- Device must be operated by using a thumb and one finger only
- Device cannot be attached to a hand or any fingers and cannot touch other parts of the hand
- Use only the materials provided

Engineering Design Process

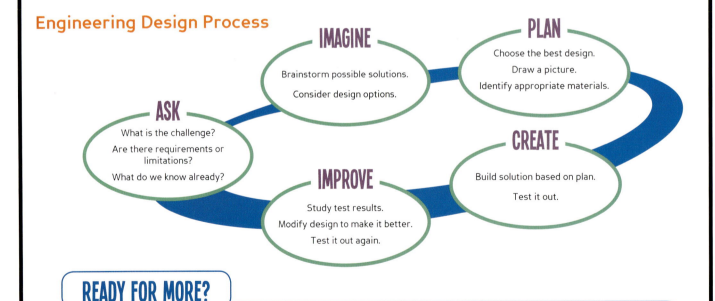

IMAGINE
Brainstorm possible solutions.
Consider design options.

PLAN
Choose the best design.
Draw a picture.
Identify appropriate materials.

ASK
What is the challenge?
Are there requirements or limitations?
What do we know already?

CREATE
Build solution based on plan.
Test it out.

IMPROVE
Study test results.
Modify design to make it better.
Test it out again.

READY FOR MORE?

If time allows, accept a new challenge for your device. Can it pick up a flat piece of paper or a cup (not using the rim)? How about something heavy, like a cell phone or a roll of masking tape? Do you need to modify your design to accomplish these new tasks?

FAMILY engineering

HOT CHOCOLATE MACHINE

Design Challenge

Design a gravity flow machine using stacked paper cups that mix water, chocolate powder, and milk powder to make a cup of hot chocolate.

Follow the engineering design process to design and test a hot chocolate machine.

Hot Chocolate Machine Design Requirements

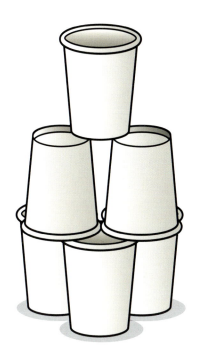

- ▶ Machine uses no more than 20 paper cups
- ▶ Machine is self-supporting (you can't use your hands to hold it together)
- ▶ There is no limit on the number and size of holes in the cups
- ▶ Chocolate and milk powders cannot be pre-mixed and must start in two separate parts of machine
- ▶ Machine works by gravity—pouring ½ cup of water into the top cup and allowing it to flow into other cups below

Engineering Design Process

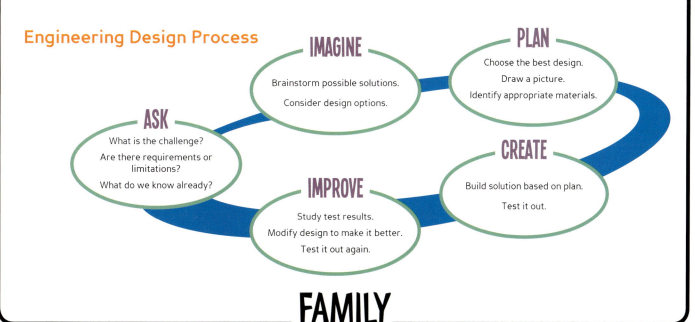

IMAGINE
Brainstorm possible solutions.
Consider design options.

PLAN
Choose the best design.
Draw a picture.
Identify appropriate materials.

ASK
What is the challenge?
Are there requirements or limitations?
What do we know already?

CREATE
Build solution based on plan.
Test it out.

IMPROVE
Study test results.
Modify design to make it better.
Test it out again.

LAUNCHER

Design Challenge

Design a launcher that can propel a cotton ball the farthest distance.

Follow the engineering design process to design and test a launcher.

Launcher Design Requirements

▶ Use only the provided materials to build your launcher

▶ Launcher may be held in place with your hands if needed

▶ You cannot use the throwing motion of your arm to operate the launcher

▶ You cannot modify or add materials to the cotton ball

Engineering Design Process

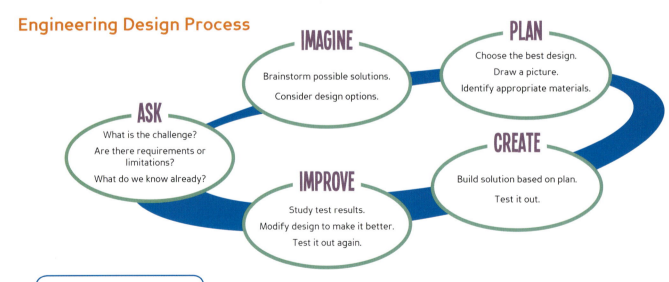

IMAGINE
Brainstorm possible solutions.
Consider design options.

PLAN
Choose the best design.
Draw a picture.
Identify appropriate materials.

ASK
What is the challenge?
Are there requirements or limitations?
What do we know already?

CREATE
Build solution based on plan.
Test it out.

IMPROVE
Study test results.
Modify design to make it better.
Test it out again.

READY FOR MORE?

- Accept a new challenge for your launcher. Can you hit a specific target on the floor? Try both close up and distant targets.

- Check out other launcher designs around the room. How are they different from yours?

FAMILY
engineering

MINING FOR CHOCOLATE

Design Challenge

> Design a process for mining chocolate chips out of a cookie to get the most chocolate chips while causing the least amount of damage to the cookie.

Mining Process Design Requirements

▶ Use only the tools provided

▶ You will have exactly 10 minutes for your extraction

▶ Whole chocolate chips are worth more than broken chocolate chips

▶ Whole cookies cost less to repair than broken or crumbled cookies

Data Sheet

	One Soft Cookie	One Hard Cookie
1. Tool(s) selected for mining		
2. How many **whole** chocolate chips we think we can extract in 10 minutes		
3. Number of **whole** chocolate chips actually extracted over 10 minutes		
4. Number of chocolate chip **pieces** extracted		
5. Value of **all** chocolate chips extracted, both whole and pieces (use chart on back side of page)		
6. Cost of environmental repair (use chart on back side of page)		
7. Mining Net Profit (subtract #6 from #5)		

MINING FOR CHOCOLATE

Charts

(use these charts to complete the data sheet on the other side)

Value of Chocolate Chips:

Whole chocolate chip	$1,000
Chocolate chip pieces	$100 each piece

Cost of Environmental Repair

Condition of the Cookie	Number of Chips to Set Aside	Cost
Cookie still whole, only pitted	1 whole chip or 10 chocolate chip pieces	$1,000
Cookie in 2 pieces	2 whole chips or 20 chocolate chip pieces	$2,000
Cookie in 3 pieces	3 whole chips or 30 chocolate chip pieces	$3,000
Cookie in 4 pieces	4 whole chips or 40 chocolate chip pieces	$4,000
Cookie in more than 4 pieces	5 whole chips or 50 chocolate chip pieces	$5,000

FAMILY engineering

ENGINEERING CHALLENGE ACTIVITY INTERACTIVES

ASSEMBLY LINE
Data Sheet

	Completion Time
Individual Assembly	
Assembly Line Trial #1	
Assembly Line Trial #2	

ASSEMBLY LINE
Data Sheet

	Completion Time
Individual Assembly	
Assembly Line Trial #1	
Assembly Line Trial #2	

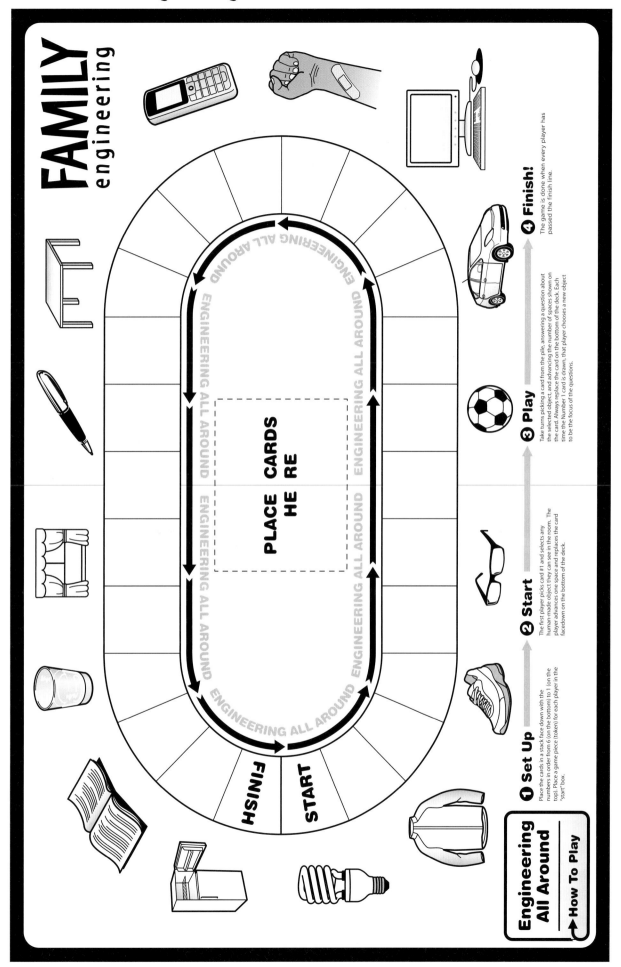

FAMILY engineering

PLACE CARDS HE RE

ENGINEERING ALL AROUND

FINISH

START

Engineering All Around

→ **How To Play**

❶ **Set Up**

Place the cards in a stack face down with the numbers in order from 6 (on the bottom) to 1 (on the top). Place a game piece (token) for each player in the "start" box.

❷ **Start**

The first player picks card #1 and selects any human-made object they can see in the room. The player advances one space and replaces the card facedown on the bottom of the deck.

❸ **Play**

Take turns picking a card from the pile, answering a question about the selected object, and advancing the number of spaces shown on the card. Always replace the card on the bottom of the deck. Each time the Number 1 card is drawn, that player chooses a new object to be the focus of the questions.

❹ **Finish!**

The game is done when every player has passed the finish line.

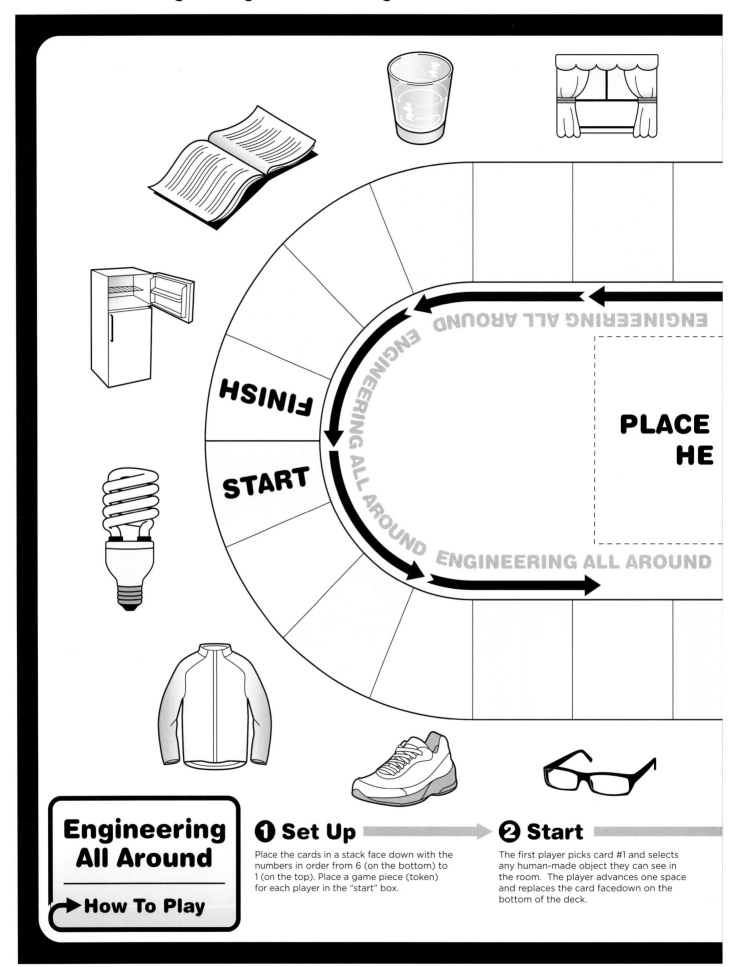

FINISH

START

ENGINEERING ALL AROUND

ENGINEERING ALL AROUND

ENGINEERING ALL AROUND

PLACE
HE

Engineering All Around

→ How To Play

❶ Set Up

Place the cards in a stack face down with the numbers in order from 6 (on the bottom) to 1 (on the top). Place a game piece (token) for each player in the "start" box.

❷ Start

The first player picks card #1 and selects any human-made object they can see in the room. The player advances one space and replaces the card facedown on the bottom of the deck.

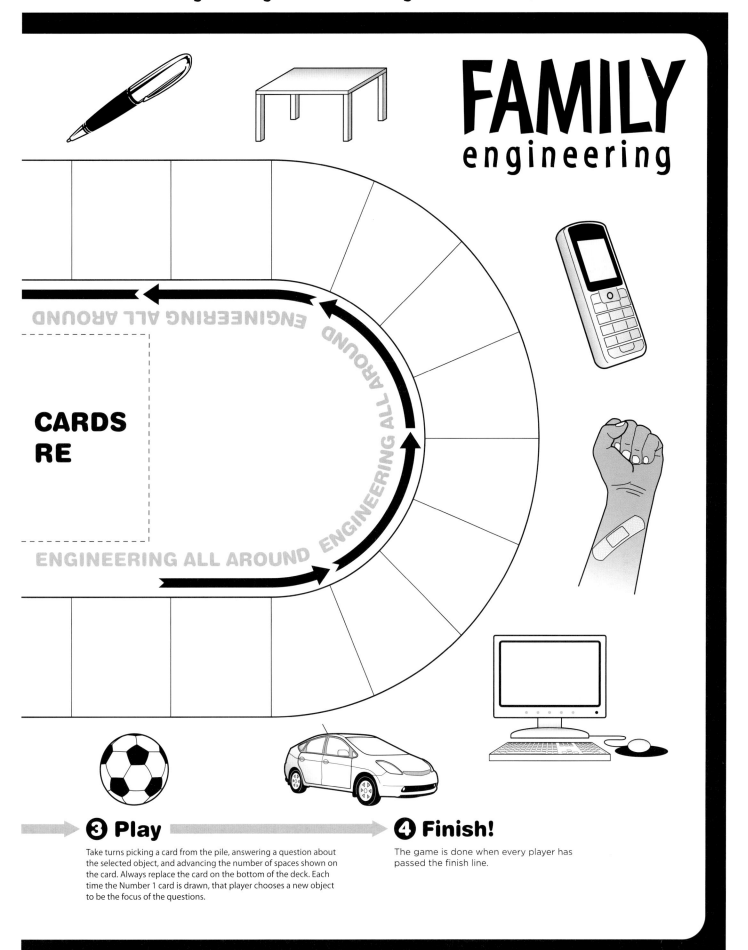

FAMILY engineering

CARDS
RE

ENGINEERING ALL AROUND

ENGINEERING ALL AROUND

❸ Play

Take turns picking a card from the pile, answering a question about the selected object, and advancing the number of spaces shown on the card. Always replace the card on the bottom of the deck. Each time the Number 1 card is drawn, that player chooses a new object to be the focus of the questions.

❹ Finish!

The game is done when every player has passed the finish line.

Engineering All Around Cards

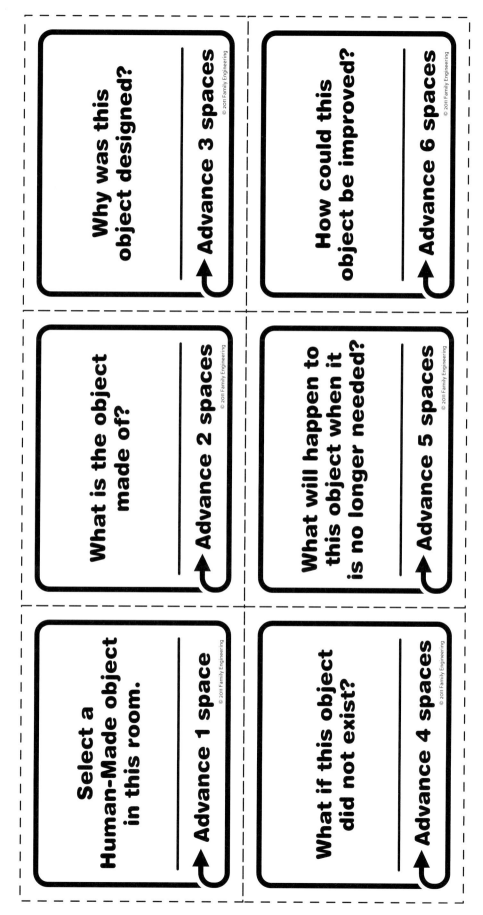

Why was this object designed?

Advance 3 spaces

© 2011 Family Engineering

How could this object be improved?

Advance 6 spaces

© 2011 Family Engineering

What is the object made of?

Advance 2 spaces

© 2011 Family Engineering

What will happen to this object when it is no longer needed?

Advance 5 spaces

© 2011 Family Engineering

Select a Human-Made object in this room.

Advance 1 space

© 2011 Family Engineering

What if this object did not exist?

Advance 4 spaces

© 2011 Family Engineering

✂ *Copy back-to-back. Cut on the dotted lines to create the Game Cards.*

✂

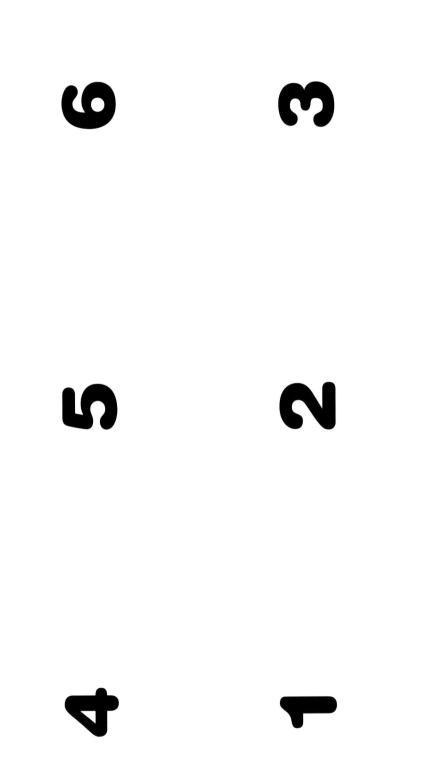

6   3

5   2

4   1

✂ Copy back-to-back. Cut on the dotted lines to create the Game Cards.

ENGINEERING CHARADES
Game Instructions

1. Each family draws one card out of the envelope. Without letting the other families in your group see the card, spend 2-3 minutes planning your charade.

2. Take turns working as a team to act out your engineered object, starting by announcing out loud the engineering field at the top of the card. The other families in your group will try to guess the object.

3. Each family has two minutes to act out their product. Have a person from another family in the group keep time on their watch or a wall clock.

4. Once all families in the group have acted out their first charade, each family chooses another card and starts the game over.

RULES FOR ENGINEERING CHARADES

*   Use only the engineered objects on the cards in the envelope.
*   Making sounds is okay, but you cannot use words to describe the object.
*   Do not point to objects in the room for hints.

MECHANICAL ENGINEER

Conveyor Belt

MATERIALS ENGINEER

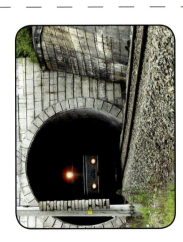

Football Helmet

MATERIALS ENGINEER

Waterproof Fabric

COMPUTER ENGINEER

Computer Mouse

BIOMEDICAL ENGINEER

Wheelchair

BIOMEDICAL ENGINEER

X-Ray Machine

MECHANICAL ENGINEER

Bicycle

GEOLOGICAL ENGINEER

Tunnel

✂ *Copy single-sided. Cut on the dotted lines to create the Engineering Charades Cards.* © 2011 Family Engineering

Engineering Charades Cards

CHEMICAL ENGINEER

Cleaning Fluid

ENVIRONMENTAL ENGINEER

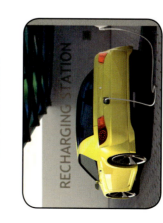

Clean Water

CIVIL ENGINEER

Skyscraper

AEROSPACE ENGINEER

Airplane

AEROSPACE ENGINEER

Rocket

CIVIL ENGINEER

Bridge

CHEMICAL ENGINEER

Post-It (Sticky) Note

MECHANICAL & ELECTRICAL ENGINEER

Electric Car

Copy single-sided. Cut on the dotted lines to create the Engineering Charades Cards.

Five Points Traffic Jam: Traffic Management Tools

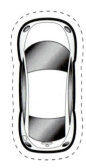 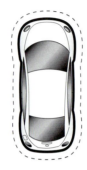  

Five Points Traffic Jam: Traffic Management Tools

Five Points Traffic Jam: Small Intersection Map

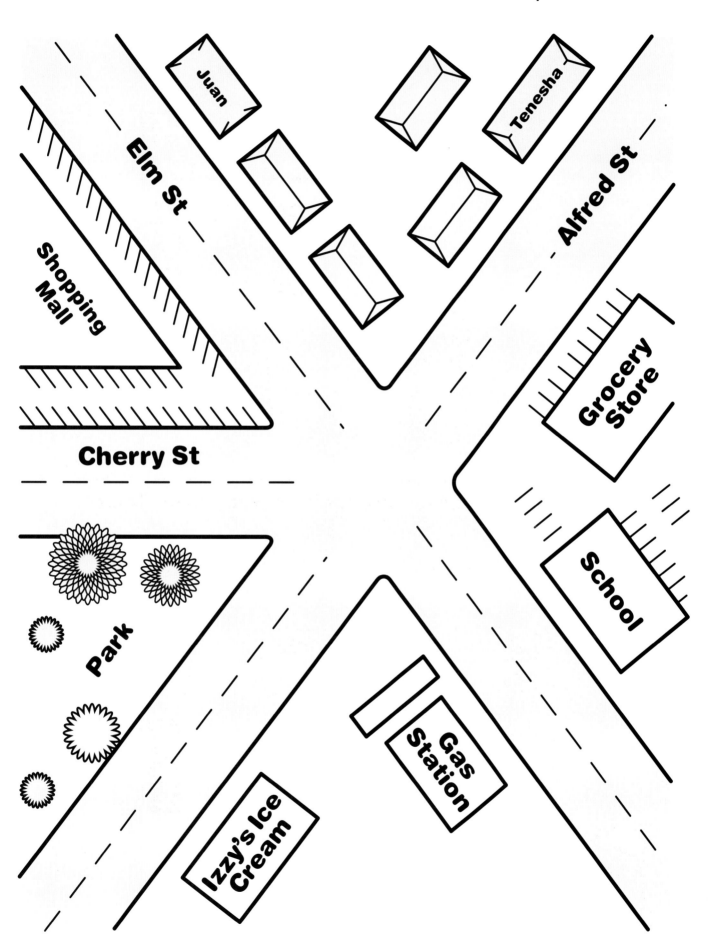

Juan

Tenesha

Elm St

Alfred St

Shopping Mall

Grocery Store

Cherry St

School

Park

Gas Station

Izzy's Ice Cream

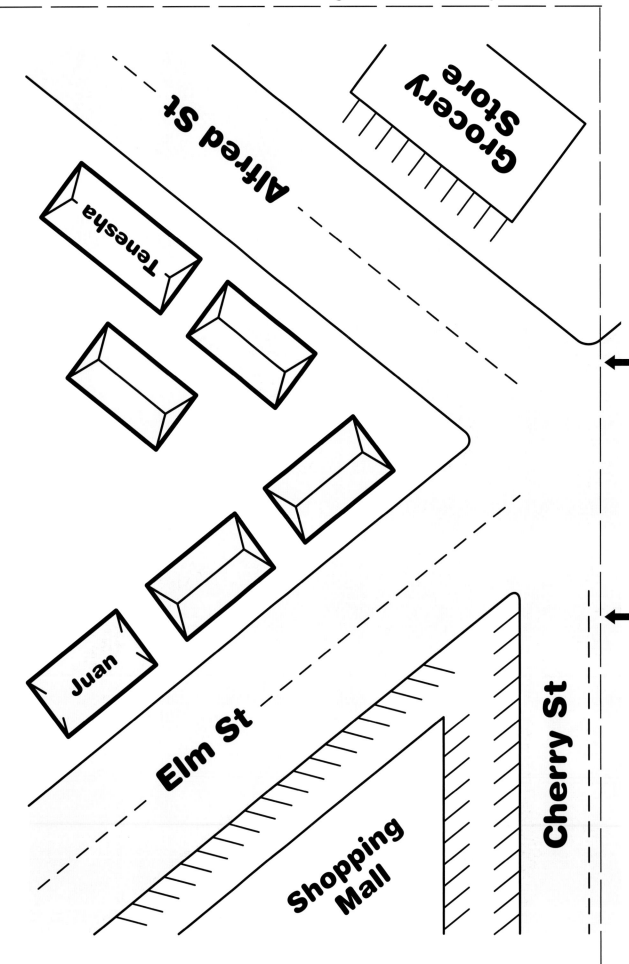

Alfred St

Grocery Store

Tenesha

Juan

Elm St

Cherry St

Shopping Mall

Cut along the dashed line. Tape Side A to Side B, lining up the Side A edge with the dashed line on Side B.

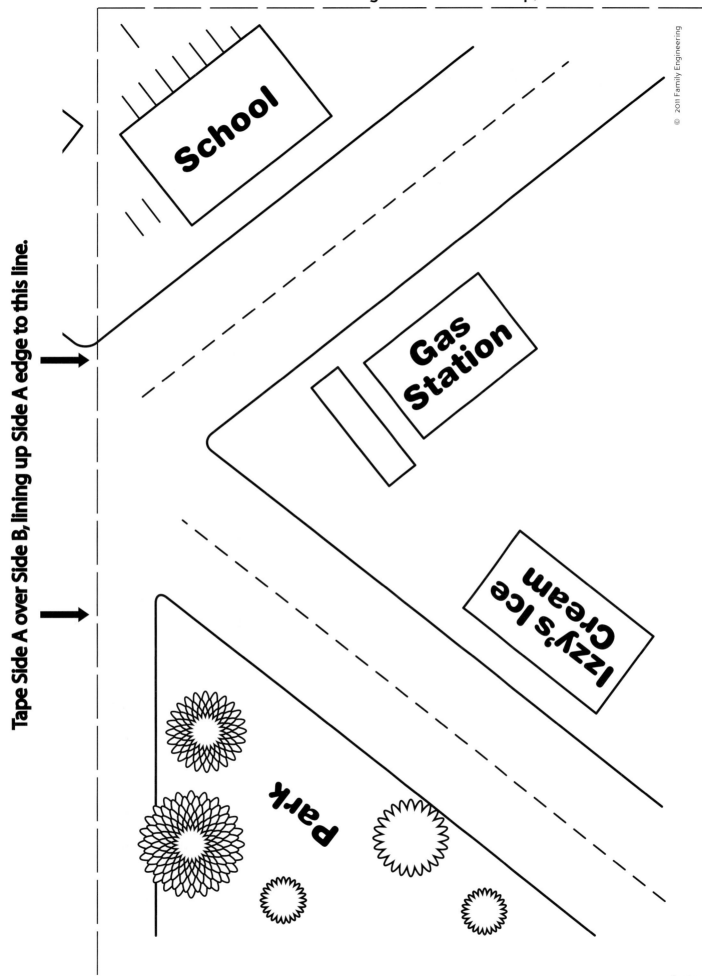

Tape Side A over Side B, lining up Side A edge to this line.

School

Gas Station

Izzy's Ice Cream

Park

STOP AND THINK
Questions

1. For what purpose or use did engineers design this object?

2. What design features help make this object work?
   (Think about materials, shape, etc.)

3. What other uses could this object have?

© 2011 Family Engineering

- -

STOP AND THINK
Questions

1. For what purpose or use did engineers design this object?

2. What design features help make this object work?
   (Think about materials, shape, etc.)

3. What other uses could this object have?

© 2011 Family Engineering
A-103

EVENT PLANNING RESOURCES

FAMILY engineering

Engineering Design Process

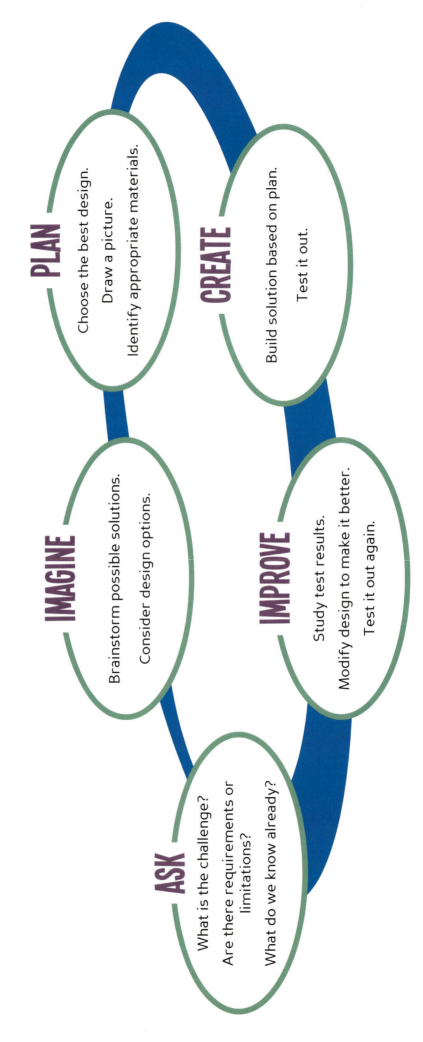

PLAN
Choose the best design.
Draw a picture.
Identify appropriate materials.

CREATE
Build solution based on plan.
Test it out.

IMAGINE
Brainstorm possible solutions.
Consider design options.

IMPROVE
Study test results.
Modify design to make it better.
Test it out again.

ASK
What is the challenge?
Are there requirements or limitations?
What do we know already?

Adapted from *Engineering is Elementary*® (EIE).
Museum of Science, Boston, MA.

www.familyengineering.org

Exploring Engineering at Home:
Tips for Parents and Other Adult Caregivers

Encourage and support an interest in engineering

By showing an interest yourself, you'll build a positive attitude toward engineering in your children. Help them explore engineering on the Internet, find books or magazines about science and engineering at the library, watch movies or television programs that involve design and engineering, or try their hand at designing and building an invention of their own. Try out some activities from *Family Engineering* at home.

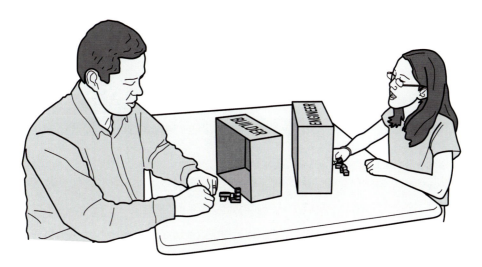

Model problem-solving strategies

Encourage children to find problems that need solving or ways of doing things that could be improved. Show them different strategies for solving problems:

- draw a picture or diagram
- talk it over with a friend or family member
- find an expert to offer advice
- break the problem down into smaller pieces
- brainstorm multiple options or approaches to a solution
- design and build a prototype for testing

Find opportunities to solve problems together as a family. Recognize when a solution or a design fails. Discuss what didn't work and how to use this experience to learn how to improve the solution.

Encourage questions

Teach children how to seek answers to their questions by reading books, using the Internet, experimenting, or taking things apart. Don't worry if you don't know the answers. Enjoy seeking out an answer with your child. Encourage children to ask good questions by posing open-ended questions yourself.

- *What happened?*
- *What should we try next?*
- *What will happen if?*
- *Can you show/tell me how this works?*

Help children "do it themselves"

Try not to simply give your children a solution or answer to a problem. Instead, encourage them to keep trying to find their own solution. Guide them with questions, hints, or clues. Let them make mistakes and correct themselves. Children who learn to work their way through problems, explore different solutions, and who learn to overcome their fear of failure become more capable and confident learners.

Support science and mathematics learning

Model a "can-do" attitude about science and math, recognizing the ways they are useful in daily life, such as counting change, predicting the weather, cooking, and fixing a flat tire on a bike. Help children develop good study habits and show an interest in what they are learning in school. Encourage participation in science and math activities outside of school, such as clubs, competitions, after-school programs, or museum classes.

Recognize the creative side of engineering

Creativity is important in engineering. Encourage your children's imaginations by providing open-ended toys, creative materials, and a safe place that allows them to design their own experiences, dream up imaginary worlds, invent interesting products, or design innovative structures.

Demonstrate how engineering improves the way we live

Take time with your child to look around and see how many items you can find that were designed by engineers. Help children see how engineered products are a part of everyday life, and that engineers design things that improve the way we live.

Challenge stereotypes about who does engineering

Find examples of female engineers or engineers from diverse cultures and backgrounds on the Internet, on television, or in your community and share these role models with your children.

www.familyengineering.org

Event Feedback Form

We are always interested in improving our Family Engineering events.
Please circle the number that best describes your family's response to each of the statements below:

		Strongly Disagree	Disagree	Neutral	Agree	Strongly Agree
1.	Our family enjoyed the event.	1	2	3	4	5
2.	The event encouraged our family to work together.	1	2	3	4	5
3.	We have a better understanding of how engineering impacts our daily lives.	1	2	3	4	5
4.	We would recommend Family Engineering to friends.	1	2	3	4	5

5. Share something your family learned about engineering.

6. How could the Family Engineering event be improved?

7. Anything else you want us to know?

Write additional comments on the back. Thank You!

© 2011 Family Engineering

Event Feedback Form

We are always interested in improving our Family Engineering events.
Please circle the number that best describes your family's response to each of the statements below:

		Strongly Disagree	Disagree	Neutral	Agree	Strongly Agree
1.	Our family enjoyed the event.	1	2	3	4	5
2.	The event encouraged our family to work together.	1	2	3	4	5
3.	We have a better understanding of how engineering impacts our daily lives.	1	2	3	4	5
4.	We would recommend Family Engineering to friends.	1	2	3	4	5

5. Share something your family learned about engineering.

6. How could the Family Engineering event be improved?

7. Anything else you want us to know?

Write additional comments on the back. Thank You!

© 2011 Family Engineering
A-109

Please join us for

FAMILY
engineering

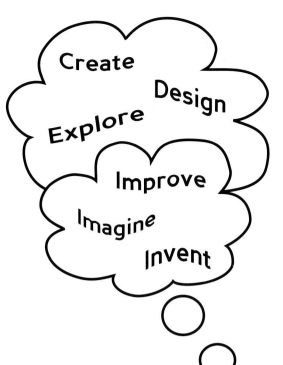

Create
Design
Explore
Improve
Imagine
Invent

Hosted by:

Discover the fun of engineering through hands-on activities for the whole family!

Date: **Time:**

Place:

- ◆ **Children ages 7-12 and their families explore engineering together!**
- ◆ **Have fun with your family!**
- ◆ **Free admission. Snacks!**

For more information, contact: _____

To register, fill in the form below and return to:

_____ by _____

- -

Yes! Sign us up for Family Engineering!

Last Name: _____

Number of Adults: _____

Number of Children: _____   Ages of Children: _____ _____ _____ _____

Phone Number: _____   Email: _____

8-12 Weeks Ahead

- ☐ Assemble a planning committee
- ☐ Identify target audience
- ☐ Identify and secure partners and contacts necessary to reach this audience
- ☐ Identify and secure an event location
- ☐ Set event date and time—reserve rooms/space at event location
- ☐ Determine maximum number of participants event space will accommodate

4-8 Weeks Ahead

- ☐ Recruit volunteers and facilitators or schedule staff to help host the event
- ☐ Set date for a training session
- ☐ Invite engineers or engineering students to participate as career role models, volunteers, or activity facilitators
- ☐ Choose activities for event and try in advance or assign to activity facilitators
- ☐ Prepare event schedule
- ☐ Develop plan for registering families for the event—printed registration forms, call-in phone number, email, etc. Registration information should include family name, number of adults and children expected to attend, ages of children, and contact information for follow-up reminder emails or phone calls.
- ☐ Create flyer for promoting the event, with detachable form, contact phone number, or email for family registration (see *Family Engineering Event Flyer* in Appendix E)
- ☐ Conduct other event advertising such as newsletters, email blasts, posters, etc.

3-4 Weeks Ahead

- ☐ Distribute flyers to target audience
- ☐ Confirm volunteers/staff for training session and event
- ☐ Arrange childcare (if needed)
- ☐ Secure raffle prizes (if needed)

1-2 Weeks Ahead

- ☐ Finalize event schedule
- ☐ Confirm engineers/engineering students as career role models
- ☐ Conduct volunteer/staff training session. Include engineers/engineering students if available. Distribute the *Working With Families: Tips for Event Volunteers* to all volunteers/staff.
- ☐ Monitor registration numbers and conduct additional promotion if needed
- ☐ Confirm details with event location—event schedule, room set-up, sound system, trash containers, after hours access if necessary
- ☐ Organize and prepare all materials/supplies

Week of the Event

- ☐ Send reminders or place follow-up calls to pre-registered families to confirm attendance
- ☐ Confirm all volunteers/staff and engineers/engineering students career role models
- ☐ Prepare nametags for volunteers/staff and engineer/engineering student career role models
- ☐ Purchase/gather refreshments and necessary serving supplies
- ☐ Make welcome signs, directional signs, and/or event programs (if needed)
- ☐ Make copies of handouts, activity signs, activity cards, and evaluation forms
- ☐ Double check materials based on estimated audience size (always prepare for extra)
- ☐ Prepare welcome and closing comments
- ☐ Invite the media (local newspaper or television station)

Day of Event

- ☐ Use a detailed event schedule for organizing and monitoring set up, volunteer/staff orientation, event activities, and clean up (see *Sample Event Schedule* on pages 30-31)

After the Event

- ☐ Thank volunteers/staff and event location hosts
- ☐ Thank sponsors and/or donors
- ☐ Store re-usable materials, signs, and planning tools for use with your next event
- ☐ Congratulate yourself and your team on a job well done!

Event Materials Checklist

- ☐ Welcome signs and event programs
- ☐ Volunteer/staff nametags
- ☐ Additional *Working With Families* handouts (for volunteers/staff)
- ☐ Blank nametags and markers for event participants
- ☐ List of registered participants
- ☐ Sign-in sheet and pencils
- ☐ Directional signs as needed
- ☐ Refreshments
- ☐ Activity handouts
- ☐ Activity materials
- ☐ *Exploring Engineering at Home* handout (for families)
- ☐ *Family Engineering Event Feedback Forms* and pencils (for families)

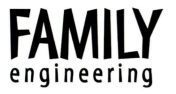

Family Name	List number of children in each grade level									Number of Adults
	PK	K	1st	2nd	3rd	4th	5th	6th	7th-8th	
1.										
2.										
3.										
4.										
5.										
6.										
7.										
8.										
9.										
10.										
11.										
12.										
13.										
14.										
15.										
16.										
17.										
18.										
19.										
20.										
21.										
22.										
23.										
24.										
25.										

As a member of the Family Engineering event team, you play an important role in engaging families in a positive and meaningful experience with engineering. Here are some tips and guidelines to help you make engineering accessible, interesting, and fun for families.

Be welcoming and friendly
Smile and interact with families to help them feel relaxed and comfortable.

Promote family interaction and involvement
Invite the whole family to work together by encouraging parents to join their children in doing the activities. Watch out for adults who "take over" during an activity and encourage families to involve all family members.

Allow families to explore and discover on their own
Try not to show or tell families how to complete an activity. Use encouragement, questions, and hints to help them explore on their own.

Ask questions
If families are stuck, confused, or just need a little encouragement, ask questions to get them back on track. Spark their creativity, help them problem-solve, or try a new approach by using the following questions.

- *What do you think will happen?*
- *What is happening?*
- *What could you do next?*
- *What is the problem you are trying to solve? What is your goal?*
- *Is there another way to think about this problem?*
- *What information do you have? What additional information do you need?*
- *How could you change your design?*
- *Would you like to try it another way?*
- *Would you like to explore how other families are approaching this challenge?*

Be enthusiastic and positive about engineering
Promote engineering as an exciting field that everyone can explore.

Make connections to engineering
Help families recognize how the activities connect engineering to their everyday lives. Point out when they are doing things that engineers do, such as problem-solving or using the engineering design process.

Share your experiences with engineering
Share with families how your own life is involved with engineering. If you are an engineer, what got you interested in engineering, what do you do in your daily work, and what do you enjoy about your job? If you are not an engineer, how do you interact with engineering and why do you think engineering is important?

Enjoy yourself!
Relax and have fun interacting with the families.

Thanks for helping to make the Family Engineering event a success!
www.familyengineering.org

Working With Families:
Tips for Engineers

A Family Engineering event is a great opportunity to inspire children and their parents and increase their interest in engineering and engineering careers. As a role model, you can bring engineering to life for families and help them see engineering as a rewarding and accessible career option. During the event, talk to families about what you do and what you like about engineering. Ask them about their own interests and, if possible, make personal connections to engineering. Most importantly, have fun and help families have a positive experience with engineering!

Messages to keep in mind while interacting with families

- Engineers are creative problem-solvers and contribute to society in meaningful ways.
- There are many different fields of engineering.
- Engineering, and the products of engineering, are parts of everyday life.
- Communication, teamwork, creativity, and problem solving are important skills in engineering. Point out when families are using these skills.
- Point out when families are engaging in the engineering design process (Ask, Imagine, Plan, Create, Improve).

Tips for engineers serving as guest presenters/speakers at a Family Engineering event

- Provide a brief, non-technical description of your work.
- Describe how your work relates to the everyday lives of families or how your work helps others.
- Discuss what you enjoy about your work.
- Tell families how you got interested in engineering and how you prepared for your career.
- Bring along any special tools, equipment, or interesting attire used in your work.
- Limit your comments to 5-10 minutes. Allow time for a few questions from families.

Frequently asked career questions

- What is it like to be a _____ ?
- Can you describe a typical day at your job?
- How much do you make? (Providing a range for starting or mid-career salaries is best.)
- How much schooling is required to become an engineer? What classes helped prepare you for a career in engineering? (math, science, writing, public speaking, computer science)
- What advice would you give someone entering the engineering field?
- What is your favorite project/product that you have been involved with?
- Can you describe a problem you have solved or are trying to solve at work?
- How does your job reward you? (opportunities to learn, travel, helping others, etc.)
- What skills besides engineering are important to your work? (communication, teamwork, creativity)
- When you were young, what did you like to do that stimulated an interest in engineering? (design/build things, take things apart)
- What did you learn in elementary school that is useful in your engineering career?
- What do you do in your free time? What are your hobbies?

Thanks for helping to make the Family Engineering event a success!
www.familyengineering.org

Photo Credits

Key

t = top	r = right	FS = Fotosearch.com
m = middle	l = left	SS = Shutterstock.com
b = bottom		SP = iStockphoto.com
		WM = Wikimedia Commons

Chapter 6

44bl, mevans/SP; 44br, MortonPhotographic/SP; 45, ez_thug/SP; 47, ACMPhoto/SP; 48, monkeybusinessimages/SP; 49, JulNichols/SP; 50bl, egdigital/SP; 50br, xyno/SP; 57, Jack11Poland/WM; 53bl, FS; 53br, Liftarn/WM; 54bl, goldenKB/SP; 54bc, Mlenny/SP; 54br, Tuomas Vitikainen/WM; 55, Maridav/SP; 56bl, jx99/SP; 56br, mevans/SP; 61, skynesher/SB; 62, Darren Baker/SS; 65, EricVega/SP; 66, mollypix/SP; 67m, jeffdalt/SP; 67tr, Lya_Cattel/SP; 69bl, TeamCrucillo/SP; 69br, LUX_8/SP; 71tl, ARENA Creative/SS; 71tr, kenneth-cheung/SP; 73, rappensuncle/SP; 74, WillMcC/WM; 75, © www.vintage-views.com; 78l-r1, fotoVoyager/SP; 78l-r2, dto3mbb/SP; 78l-r3, lisafx/SP; 78l-r4, ronen/SP; 78l-r5, endopack/SP; 78l-r6, Leah-Anne Thompson/SS; 81, ideabug/SP; 82, Photomorphic/SP

Chapter 7

87, Atelier Pictures; 88, James Steidl/SS; 89, used by permission of Sony Electronics, Inc.; 91, Atelier Pictures; 92, Public Domain/WM; 96, christimatei/SP; 97, perkijl/SP; 100, gnagel/SP; 101, Snell Memorial Foundation; 103, Public Domain/WM; 107, © Werner Forman/Art Resource, NY; 108, Reprinted with the permission of Little Simon, an imprint of Simon & Schuster Children's Publishing Division from ALICE'S ADVENTURES IN WONDERLAND A Pop-Up Adaptation by Robert Sabuda. Copyright © 2003 Robert Sabuda; 109, Atelier Pictures; 110, Atelier Pictures; 111, Atelier Pictures; 113, Kiyyah/SP; 114l-r1, thelinke/SP; 114l-r2, MAridav/SP; 114l-r3, ChrisMR/SP; 114l-r4t, yumiyum/SP; 114l-r4b, LSOphoto/SP; 117, Tony Campbell/SS; 118, WM; 121, used by permission of Touch Bionics; 124, XuRa/SS; 125t, Atelier Pictures; 125b, boesephoto/SP; 127, used by permission of World Championship Pumpkin Chunkin; 128t, Public Domain/WM; Harrieta171/WM; 129, oliveromg/SS; 131, Public Domain/WM; 132tr, BartCo/SP; 132bl, Mabus13/SP; 132br, © www.airphotana.com; 133, © www.airphotana.com; 136, iofoto/SP; 139t, Public Domain/WM; 139b, Public Domain/WM

Appendix A

A-4bl, mevans/SP; A-4br, MortonPhotographic/SP; A-6tl, GA161076; A-6bl, Marbury/SP; A-6br, ez_thug/SP; A-8bl, Public Domain/WM; A-8br, ACMPhoto/SP; A-10, JulNichols/SP; A-12tr, egdigital/SP; A-12mr, xyno/SP; A-12bl, Jack11Poland/WM; A-14mr, Liftarn/WM; A-14bl, FS; A-16, Maridav/SP; A-22, skynesher/SP; A-24, Darren Baker/SS; A-26, EricVega/SP; A-28bl, mollypix/SP; A-28br, Lya_Cattel/SP; A-30tr, LUX_8/SP; A-30mr, TeamCrucillo/SP; A-32tr, ARENA Create/SS; A-32bl, kenneth-cheung/SP; A-34, rappensuncle/SP; A-35l-r1, mjbs/SP; A-35l-r2, WillMcC/WM; A-35l-r3, Rouzes/SP; A-36tr, Erechtheus/WM; A-36, © www.vintage-views.com; A-38tr, BirdImages/SP; A-38bl, dsharpie/SP; A-38br, mdecoste/SP; A-42l-r1, fotoVoyager/SP; A-42l-r2, dto3mbb/SP; A-42l-r3, lisafx/SP; A-42l-r4, ronen/SP; A-42l-r5, endopack/SP; A-42l-r6, Leah-Anne Thompson/SS; A-42, ideabug/SP; A-46, Photomorphic/SP

Appendix B

A-52tl, Tuomas Vitikainen/WM; A-52tr, mathiaswilson/SP; A-52bl, kate_sept2004/SP; A-52br, Public Domain/WD; A-53tl, istockKB/SP; A-53trhockeymom4/SB; A-53bl, Piepereit/SB; A-53br, jorgeantonio/SP; A-54tl, Mia Jackson; A-54tr Maridav/SP; A-54bl, Mlenny/SP; A-54br, Karma_pema/SP; A-56tl1, neukind/SP; A-56tl2, stocksnapper/SP; A-56tr, design56/SP; A-56bl1, cgering/SP; A-56bl2, rubenhi/SP; A-56br, technotr/SP; A-57tl, Public Domain/WM; A-57tr, jx99/SP; A-57bl1, David Brazier/WM; A-57bl2 chrisho/SP; A-57br, Ryan Somma/WM; A-59tl, manu-manoun/SP; A-59tr, STILLFX/SS; A-59bl, renaschild/SP; A-59br, uwimages/SP; A-61tl, Andyworks/SP; A-61tr, mevans/SP; A-61bl, J Brew/WM; A-61br, Kristof Degreef/SS; A-69tl, Leah-Anne Thompson/SS; A-69tr, Blend_Images/SP; A-69bl, darrenwise/SP; A-69br, runen/SP; A-70tl, lisafx/SP; A-70tr, racheldonahue/SP; A-70bl, endopack/SP; A-70br, dto3mbb/SP; A-71tl, carlosalvarez/SP; A-71tr, sjlocke/SP; A-71bl, Mari/SP; A-71br, fotoVoyager/SP; A-73tl, bakalusha/SP; A-73tr, Kameleon007/SP; A-73bl, monkeybusinessimages/SP; A-73br, Michael Svoboda/SP

Appendix C

A-80, Atelier Pictures; A-82, used by permission of Touch Bionics

Appendix D

A-96 (bicycle), Carlosavarez/SP; A-96 (wheelchair), hidesy/SP; A-96 (waterproof fabric), Maridav/SP; A-96 (conveyor belt), thelinke/SP; A-96 (tunnel), Lucian Cretu/SS; A-96 (x-ray machine), LSOphoto/SP; A-96 (x-ray), yumiyum/SP; A-96 (computer mouse), bc173/SP; A-96 (football helmet), aida_b/SP; A-97 (post-it note), Kursad/SP; A-97 (rocket), Public Domain/WM; A-97 (skyscraper), Emily Rivera/SP; A-97 (cleaning fluid), laylandmasuda/SP; A-97 (electric car), SS; A-97 (bridge), ChrisMR/SP; A-97 (airplane), egdigital/SP; A-97 (clean water), IJzendoorn/SP; A-102tl, ChrisO/WM; A-102tr, Public Domain/WM; A-102ml, oliveromg/SS; A-102mr, used by permission of World Championship Pumpkin Chunkin; A-102bl1, Driph/WM; A-102bl2, Lobsterclaws/SP; A-102br, Bobby Deal/RealDealPhoto/SS